CAMBRIDGE LIBRARY COLLECTION

Books of enduring scholarly value

Botany and Horticulture

Until the nineteenth century, the investigation of natural phenomena, plants and animals was considered either the preserve of elite scholars or a pastime for the leisured upper classes. As increasing academic rigour and systematisation was brought to the study of 'natural history', its subdisciplines were adopted into university curricula, and learned societies (such as the Royal Horticultural Society, founded in 1804) were established to support research in these areas. A related development was strong enthusiasm for exotic garden plants, which resulted in plant collecting expeditions to every corner of the globe, sometimes with tragic consequences. This series includes accounts of some of those expeditions, detailed reference works on the flora of different regions, and practical advice for amateur and professional gardeners.

Elements of Botany

Employed early in his career by Sir Joseph Banks, the botanist John Lindley (1799–1865) is best known for his recommendation that Kew Gardens should become a national botanical institution, and for saving the Royal Horticultural Society from financial disaster. As an author, he is best remembered for his works on taxonomy and classification. A partisan of the 'natural' system rather than the Linnaean, Lindley published this 1841 work, the fourth edition of his *Outline of the First Principles of Botany*, under a new title, to emphasise not only that it was 'much extended, and, it is hoped, improved', but also that it was a textbook for students of 'structural, physiological, systematical, and medical' botany. He defines the different elements of a plant, and provides a checklist for identification of plant families, before discussing the various 'natural' systems of classification, including his own, and the different practical uses of plants.

Cambridge University Press has long been a pioneer in the reissuing of out-of-print titles from its own backlist, producing digital reprints of books that are still sought after by scholars and students but could not be reprinted economically using traditional technology. The Cambridge Library Collection extends this activity to a wider range of books which are still of importance to researchers and professionals, either for the source material they contain, or as landmarks in the history of their academic discipline.

Drawing from the world-renowned collections in the Cambridge University Library and other partner libraries, and guided by the advice of experts in each subject area, Cambridge University Press is using state-of-the-art scanning machines in its own Printing House to capture the content of each book selected for inclusion. The files are processed to give a consistently clear, crisp image, and the books finished to the high quality standard for which the Press is recognised around the world. The latest print-on-demand technology ensures that the books will remain available indefinitely, and that orders for single or multiple copies can quickly be supplied.

The Cambridge Library Collection brings back to life books of enduring scholarly value (including out-of-copyright works originally issued by other publishers) across a wide range of disciplines in the humanities and social sciences and in science and technology.

Elements of Botany

Structural, Physiological,
Systematical, and Medical

JOHN LINDLEY

CAMBRIDGE
UNIVERSITY PRESS

University Printing House, Cambridge, CB2 8BS, United Kingdom

Cambridge University Press is part of the University of Cambridge.
It furthers the University's mission by disseminating knowledge in the pursuit of
education, learning and research at the highest international levels of excellence.

www.cambridge.org
Information on this title: www.cambridge.org/9781108076647

© in this compilation Cambridge University Press 2015

This edition first published 1841
This digitally printed version 2015

ISBN 978-1-108-07664-7 Paperback

Selected botanical reference works available in the
CAMBRIDGE LIBRARY COLLECTION

al-Shirazi, Noureddeen Mohammed Abdullah (compiler), translated by
Francis Gladwin: *Ulfáz Udwiyeh, or the Materia Medica* (1793)
[ISBN 9781108056090]

Arber, Agnes: *Herbals: Their Origin and Evolution* (1938)
[ISBN 9781108016711]

Arber, Agnes: *Monocotyledons* (1925) [ISBN 9781108013208]

Arber, Agnes: *The Gramineae* (1934) [ISBN 9781108017312]

Arber, Agnes: *Water Plants* (1920) [ISBN 9781108017329]

Bower, F.O.: *The Ferns (Filicales)* (3 vols., 1923–8) [ISBN 9781108013192]

Candolle, Augustin Pyramus de, and Sprengel, Kurt: *Elements of the Philosophy
of Plants* (1821) [ISBN 9781108037464]

Cheeseman, Thomas Frederick: *Manual of the New Zealand Flora*
(2 vols., 1906) [ISBN 9781108037525]

Cockayne, Leonard: *The Vegetation of New Zealand* (1928)
[ISBN 9781108032384]

Cunningham, Robert O.: *Notes on the Natural History of the Strait of Magellan
and West Coast of Patagonia* (1871) [ISBN 9781108041850]

Gwynne-Vaughan, Helen: *Fungi* (1922) [ISBN 9781108013215]

Henslow, John Stevens: *A Catalogue of British Plants Arranged According to
the Natural System* (1829) [ISBN 9781108061728]

Henslow, John Stevens: *A Dictionary of Botanical Terms* (1856)
[ISBN 9781108001311]

Henslow, John Stevens: *Flora of Suffolk* (1860) [ISBN 9781108055673]

Henslow, John Stevens: *The Principles of Descriptive and Physiological Botany*
(1835) [ISBN 9781108001861]

Hogg, Robert: *The British Pomology* (1851) [ISBN 9781108039444]

Hooker, Joseph Dalton, and Thomson, Thomas: *Flora Indica* (1855)
[ISBN 9781108037495]

Hooker, Joseph Dalton: *Handbook of the New Zealand Flora* (2 vols., 1864–7) [ISBN 9781108030410]

Hooker, William Jackson: *Icones Plantarum* (10 vols., 1837–54) [ISBN 9781108039314]

Hooker, William Jackson: *Kew Gardens* (1858) [ISBN 9781108065450]

Jussieu, Adrien de, edited by J.H. Wilson: *The Elements of Botany* (1849) [ISBN 9781108037310]

Lindley, John: *Flora Medica* (1838) [ISBN 9781108038454]

Müller, Ferdinand von, edited by William Woolls: *Plants of New South Wales* (1885) [ISBN 9781108021050]

Oliver, Daniel: *First Book of Indian Botany* (1869) [ISBN 9781108055628]

Pearson, H.H.W., edited by A.C. Seward: *Gnetales* (1929) [ISBN 9781108013987]

Perring, Franklyn Hugh et al.: *A Flora of Cambridgeshire* (1964) [ISBN 9781108002400]

Sachs, Julius, edited and translated by Alfred Bennett, assisted by W.T. Thiselton Dyer: *A Text-Book of Botany* (1875) [ISBN 9781108038324]

Seward, A.C.: *Fossil Plants* (4 vols., 1898–1919) [ISBN 9781108015998]

Tansley, A.G.: *Types of British Vegetation* (1911) [ISBN 9781108045063]

Traill, Catherine Parr Strickland, illustrated by Agnes FitzGibbon Chamberlin: *Studies of Plant Life in Canada* (1885) [ISBN 9781108033756]

Tristram, Henry Baker: *The Fauna and Flora of Palestine* (1884) [ISBN 9781108042048]

Vogel, Theodore, edited by William Jackson Hooker: *Niger Flora* (1849) [ISBN 9781108030380]

West, G.S.: *Algae* (1916) [ISBN 9781108013222]

Woods, Joseph: *The Tourist's Flora* (1850) [ISBN 9781108062466]

For a complete list of titles in the Cambridge Library Collection please visit:
www.cambridge.org/features/CambridgeLibraryCollection/books.htm

ELEMENTS OF BOTANY,

STRUCTURAL, PHYSIOLOGICAL, SYSTEMATICAL, AND MEDICAL;

BEING A FOURTH EDITION OF

THE OUTLINE OF THE FIRST PRINCIPLES OF BOTANY.

BY JOHN LINDLEY, Ph.D. F.R.S.

VICE-SECRETARY OF THE HORTICULTURAL SOCIETY OF LONDON;
PROFESSOR OF BOTANY IN UNIVERSITY COLLEGE, LONDON,
THE ROYAL INSTITUTION OF GREAT BRITAIN, AND TO THE SOCIETY OF APOTHECARIES.

LONDON:

PRINTED FOR TAYLOR AND WALTON,

BOOKSELLERS AND PUBLISHERS TO UNIVERSITY COLLEGE,

UPPER GOWER STREET.

———

1841.

LONDON:

PRINTED BY SAMUEL BENTLEY,
Bangor House, Shoe Lane.

PREFACE.

The work now laid before the public is a fourth edition of the Author's "Outline of the First Principles of Botany," much extended and, it is hoped, improved. That work was written for the use of students, and entirely for the purpose of enabling them to fix correctly in their minds the more important points which the teacher brings before them in an academical course. When facts are mixed up with extended discussions, and rapidly adverted to, either in a lecture-room or in a written dissertation, the beginner is apt to lose sight of the exact nature of an argument, and is unable to distinguish with certainty the points upon which it is most material for him to fix his attention. That there existed a want of such a work has been sufficiently proved by the many editions the original Outline has passed through, in various European languages: indeed, while the present new edition is in the press, advice has been received of the translation of the work into Hungarian. The propositions which it contained were such as it is of the most indispensable importance for a student to understand; and were all, apparently, deducible from the evidence which had at that time been collected by Botanists. —The wish of the Author was to sketch a slight but accurate outline, the details of which were to be filled up by the reader himself, who, for this purpose, was referred to the Author's more extended Introductions to Botany.

The original "Outline" contained nothing more than the fundamental propositions upon which the principles of Organic and Physiological Botany depend; but, when two editions had been exhausted, the Author was induced, by the favour with which the book had been received, and by its recognized utility, notwithstanding its many defects, to combine with it a sketch of Systematical Botany, treated in the same manner. He undertook the far more difficult task of reducing to their simplest expression the characters that distinguish the various groups in which plants are classified by modern systematical writers; the object being to diminish, by a very careful and extensive analysis, the difficulties which present themselves to the student of this branch of the subject. The attempt was made in the form of a series of tables, called the "Alliances of Plants;" and it has been satisfactory to the Author to find that this too has been advantageous to students, notwithstanding its extreme conciseness. The work thus altered appeared in 1835, under the title of "Key to Structural, Physiological, and Systematical Botany."

In the edition now offered to students many important improvements have been introduced, without deviating from the original plan of the work. The skill of the wood-engraver has enabled the Author to fill his pages with illustrations, explanatory not only of the technical terms employed in Botany, but also of the Natural Orders of plants. An analysis of the latter, upon the plan of Lamarck, an account of De Candolle's celebrated system of arrangement, into which a large number of wood-cuts are introduced, and some new views relating to natural classification, are added to the matter to be found in previous editions: besides which, the whole of the Structural and Physiological part has been corrected with great care, and made to include all the most important views of modern physiologists, so as to present the reader with a view of the state of Botanical knowledge in these departments in the spring of 1841.

It is hoped that these improvements will render the work what it was originally intended for,—a complete Botanical Note-book,—wherein all the principal topics which the teachers of Botany introduce into their lectures are arranged methodically. The student will naturally look to his instructor or to more extensive works for explanations of those points which in his Note-book are merely adverted to.

University College, London,
April 1841.

ELEMENTS OF BOTANY;

STRUCTURAL, PHYSIOLOGICAL, SYSTEMATICAL,

AND MEDICAL.

I.—STRUCTURAL AND PHYSIOLOGICAL BOTANY.

---

1. PLANTS are not separable from animals by any absolute character; the simplest individuals of either kingdom not being distinguishable by our senses.

2. Animals are for the most part incapable of multiplying by mechanical or spontaneous division of their trunk, and are supported by nutritious matter, carried into their system from an internal bag or stomach.

3. Plants are for the most part congeries of individuals, multiplying by spontaneous or artificial division of their trunk or axis, and are supported by nutritious matter conveyed into their system by the absorption of their lower extremities or roots, or by their surface.

4. Generally speaking, the latter are fixed to some substance from which they grow, are destitute of locomotion, and are enabled to digest their food by the action of light upon their epidermis.

5. Plants consist of a hygrometrical membranous transparent tissue, chemically composed of oxygen, hydrogen, and carbon, to which nitrogen is always superadded. They are also found to contain many mineral substances, which they are supposed to separate from their proper food during the process of digestion, and to deposit in their tissue.

B

---

6. Their component parts are held together by an organic mucus, out of which the tissue itself is generated.

7. Tissue is found in the form of the *cellular*, the *woody*, the *vascular*, the *pitted*, and the *laticiferous*, each of which has certain modifications, constituting the Elementary organs.

I.—ELEMENTARY ORGANS.

8. Of these, CELLULAR TISSUE (Tela cellulosa, *Lat.*; Tissu cellulaire, *Fr.*; Pulp and Parenchyma, *of old writers*; Zellengewebe, *Germ.*) is the only form universally found in plants; the other forms are often either partially or entirely wanting.

9. Cellular tissue is composed of vesicles, the sides of which are not originally perforated by visible pores (22).

10. Each vesicle is a distinct individual, cohering with the vesicle with which it is in contact; and originating from a primitive point or *cytoblast*[1], which either remains visible on its sides or is absorbed.

11. Therefore the apparently simple membrane which divides two contiguous cells is in fact double.

12. If the adhesion of the contiguous cells be imperfect, spaces will exist between them. Such spaces are called *intercellular passages*.

13. The sides of cellular tissue are often thickened by the deposit, on their inner surface, of *matter of lignification* or *sclerogen*[2], which is stratified, and often pierced with passages leading to the circumference.

14. The cells contain fluid; grains of colouring matter (*chromule, chromogen,* or *chlorophyll*); starch in granules (*perenchyma*); and crystals, which, when acicular, are named *raphides*.

15. The vesicles of cellular tissue, when separate, are round or oblong; when slightly and equally pressed together, they acquire a dodecahedral appearance[4], with an hexagonal section; stretched lengthwise they become prismatical, cylindrical, fusiform, &c.

16. When cellular tissue is composed of vesicles fitting together by their plane faces, it is called in general terms *parenchyma*; and *prosenchyma* if the vesicles are fusiform. Both these are sometimes branched, and their divisions inosculate.

Spheroidal cellular tissue is *merenchyma*[4], or *sphærenchyma*; conical, *conenchyma*[5]; oval, *ovenchyma*[6]; fusiform, *atractenchyma*; cylindrical, *cylindrenchyma*[8]; sinuous, *colpenchyma*[9]; branched, *cladenchyma*[10]; prismatical, *prismenchyma*, which, when compressed, becomes *muriform*[7]; stellate, *actinenchyma*[12]; entangled, branched and tubular, *dædalenchyma*[11].

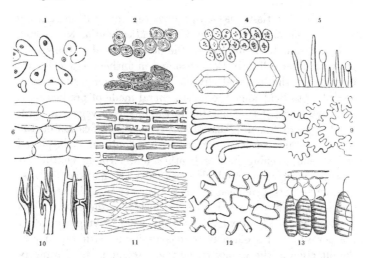

17. Parenchyma constitutes all the pulpy parts; the medulla or pith (98), the medullary rays (132), a portion of the bark (120), and all that intervenes between the veins of leaves and other appendages of the axis. Consequently it occurs in every part of a plant, and especially in those which are succulent. It, however, sometimes acquires, by the deposit of sclerogen (13), excessive hardness, as in the stone of fruits[3], and the bony skin of some seeds.

Clestines are large cells of Parenchyma, in which raphides (60 a) are often deposited.

18. Prosenchyma is confined to the bark and wood, in which it only occasionally occurs.

19. Besides these a spiral line is often found in the inside of a cell, when *fibro-cellular* tissue[13], or *inenchyma*, is produced; and it sometimes happens that the membrane connecting such fibres is absorbed, leaving the fibres only to constitute the cell.

20. The function of the cellular tissue is to transmit fluids in all directions; the membrane of which it is composed is therefore permeable, although not in general furnished with

visible pores (9). When it is thickened by the deposition of
sclerogen, passages are left in the latter communicating with
the sides of the tissue, and giving it the appearance of being
dotted or pitted.

21. Cellular tissue is self-productive, one cell generating
others upon its surface. In Chara, Marchantia, &c. young
cells are said to be formed at the points of and in the spaces
between older cells; in Confervæ and in anthers new cells are
formed by the internal divisions of an older cell; while,
according to Schleiden, the most general mode of production is
from cytoblasts (10), generated in the mucus of vegetation (6).

22. PITTED TISSUE (*Bothrenchyma*) is a modification of the
cellular, either consisting of ordinary cylindrical cells placed
end to end, opening into each other, and forming continuous
tubes; or originally tubular[15]. Its sides are marked by pits,
resembling dots, produced in consequence of the sclerogen (13)
being unequally deposited over the inside of the cells. It is
common in wood, of which it forms what is vulgarly called
the porosity. Its office is to convey fluids with rapidity in
the direction of the woody tissue that surrounds it. Formerly
it was considered a form of vascular tissue, and called *dotted
ducts*, or *vasiform tissue*.

> Pitted Tissue is articulated, when composed of short cylinders placed end to
> end, or continuous when it was originally tubular.

23. WOODY TISSUE (*Pleurenchyma*) consists of elongated
tubes tapering to each end, and, like the vesicles of cellular
tissue, imperforate to the eye. It may be considered a form of
the cellular tissue itself, to which it is frequently referred; but
it is practically distinguished by its cylindrical form, great
length, extreme fineness, and toughness; the latter of which
properties is produced by the thickness of its sides.

24. It is found in the wood, among the parenchyma of the
liber (124), and in the veins of the leaves, or other appendages
of the axis.

25. Its functions are to give strength to the vegetable fabric,
and to serve as a medium for the passage of fluid from the
lower to the upper extremities.

> Common Pleurenchyma has its sides destitute of markings; the glandular[14] is a
> variety in which the sides of the tubes are furnished with circular disks; the
> latter occur chiefly in coniferous plants and such as have aromatic secretions.

26. Vascular Tissue (*Trachenchyma*) consists of very thin-sided cylinders tapering to each end, and having a spiral fibre generated in their inside.

27. Of this kind of tissue *spiral vessels*[15][16] are the type. Their fibre is of a highly elastic nature, and is capable of unrolling when stretched.

28. Spiral vessels are found in the medullary sheath, and in all parts that emanate from it, especially the veins of the leaves, and everything that is a modification of them.

29. They are usually absent from the wood and bark. They, however, occur in these and other unusual parts in a few extremely rare cases; as in the wood, and bark, and pith of Nepenthes.

30. The spiral vessels appear intended for the conveyance of air, which has been found to contain 7 or 8 per cent. more oxygen than the atmosphere.

31. *Ducts* are transparent tubes, the sides of which are marked with rings, bars, or transverse streaks.

32. They are slight modifications of the spiral vessel, differing principally in being incapable of unrolling; and, in some cases, in the turns of the spiral fibre being distant or broken, or even, in appearance, branched.

33. In those cases where the turns of the spire actually touch each other, the ducts, which are then called *closed*, can only be distinguished from spiral vessels by their inability to unrol; while at rest they appear to be absolutely the same.

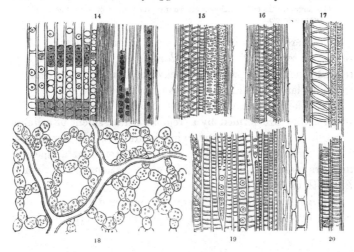

Ducts are closed[20] when the spires touch each other; annular, when they seem to consist of separate rings[17]; reticulated, when the spires cross each other[19]; scalariform, when the lines upon their sides are horizontal and equidistant; septate, when the interior is divided by pierced disks, as in Echinocactus.

34. Ducts occur among the woody tissue of herbaceous plants; are abundant in the wood of the higher tribes of cellular plants, such as Ferns and Lycopodiaceæ; and their ends are often in immediate connection with the loose cellular tissue occupying the extremities of the roots.

35. Their functions have not been accurately determined. It is probable that they act as spiral vessels when young; but it is certain that they become filled with fluid as soon as their spires are separated.

36. LATICIFEROUS TISSUE[18] (*Cinenchyma*) consists of uninterrupted anastomozing tubes, whose final divisions are so delicate, that the eye only discovers them when aided by the most powerful microscopes. It forms the *proper vessels* of old writers.

37. It principally occurs in the liber of Exogens (124), whence the ramifications proceed to the surface of all the organs, and penetrate the hairs, where they form a most delicate network.

38. Laticiferous tissue conveys *latex*, a peculiar fluid, usually turbid, and coloured red, white, or yellow; often however colourless.

39. The use of this tissue is to carry the latex to all the newly formed organs, which are supposed to be nourished by it.

The large trunks of Cinenchyma are *vasa expansa*, or *opophora*; the small are *vasa contracta*.

40. There are no other elementary forms of tissue. *Airvessels*, *Reservoirs of oil*, *Lenticular glands*, are all either distended intercellular passages, or cavities built up with cellular tissue, or large cells filled with peculiar secretions.

41. When such cavities are essential to the existence of a species, they are formed by a regular arrangement of cellular tissue in a definite and unvarying figure; *Ex*. Water-plants. When they are not essential to the existence of a species, they are mere irregular distensions or lacerations of the tissue; *Ex*. Pith of the Walnut-tree.

42. All these forms of tissue are enclosed within a skin called the epidermis, which is one or more external layers of

parenchyma, the vesicles of which are compressed, and in a firm state of cohesion.

43. The spaces seen upon the epidermis, when examined by a microscope, represent these vesicles.

44. It is, therefore, not a peculiar membrane, but a form of cellular tissue.

45. It is spread over all the parts of plants which are exposed to air, except the stigma (397).

46. It is not found upon parts habitually living under water.

47. It is itself protected by an extremely thin pellicle, which is apparently inorganic and homogeneous, and which covers every part, except the openings through the stomates (49). This membrane is the cuticle.

48. The epidermis is furnished with stomates.

49. STOMATES are oval spaces lying between the sides of the cells, opening into intercellular cavities in the subjacent tissue, and appearing to be bordered by a limb when they are viewed from above[22][23][25].

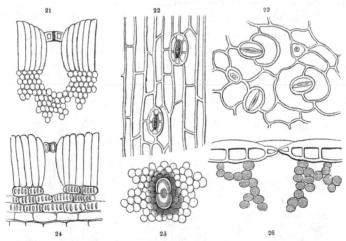

50. This appearance of a limb is owing to the juxtaposition of two or more elastic vesicles, closing up or opening the aperture which they form, according to circumstances, as is manifest when the stomate is divided perpendicularly to the plane of the epidermis [21][24][26].

51. Stomates are found abundantly upon leaves, particularly on the lower surface of those organs; occasionally upon all

parts that are modifications of leaves, especially such as are of a leafy texture; and on the stem.

52. Stomates have not been found upon the roots, nor on colourless parasitical plants, nor the submersed parts of plants, nor on Fungi, Algæ, and Lichens; they are, moreover, rare, or altogether absent, in succulent parts and in seeds.

53. It frequently happens, that they are so incompletely formed, as to be either altogether incapable of action, or to act in a very imperfect manner; as in succulent plants.

54. The function of stomates is to regulate evaporation and respiration. It has been thought, that the former function, in particular, is that for which they are destined; and, that the cause of certain parts becoming succulent, is the absence of stomates in sufficient numbers to carry off the watery part of the sap. But some succulent plants have more stomates than ordinary plants, so that this opinion requires reconsideration.

55. HAIRS are minute expansions of transparent cellular tissue proceeding from the surface of plants. They are of two kinds, lymphatic and secreting.

56. *Lymphatic* hairs are formed by vesicles of cellular tissue placed end to end, and not varying much in dimensions.

57. *Glandular* hairs are formed by vesicles of cellular tissue placed end to end, and sensibly distended at the apex or base into receptacles of fluid.

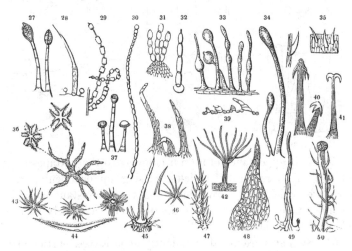

58. Lymphatic hairs are for the absorption of moisture, for the protection of the surface on which they are placed, and for the control of evaporation through the stomates (49). They always proceed from the veins, while the stomates occupy the interjacent parenchyma.

59. Glandular hairs are receptacles of the fluid peculiar to certain species of plants, such as the fragrant volatile oil of the sweet brier, and the acrid colourless fluid of the nettle, and may be regarded as organs of excretion.

Hairs are simple[35] ; setaceous[28] ; capitate[37] ; strangulated[29] ; moniliform[30] ; articulated[31][32] ; septate[27] ; compound[38] ; knotted[33] ; clavate[34] ; scabrous[39] ; ciliated[50] ; glochidiate[40][41] ; branched[42] ; stellate[45][46] ; scutate[43] ; araneose[36] ; ramentaceous[48][49].

60. Hairs are usually planted, more or less perpendicularly, upon the surface on which they grow. In some cases, however, they are attached by their middle (*peltate*), as in Malpighiaceous and Brassicaceous plants[51].

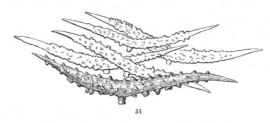

51

60 *a*. Raphides are crystals of any kind, usually acicular, found in the interior of cells of parenchyma.

61. Prickles are conical hairs of large size, sharp-pointed, and having thin tissue very hard.

II. COMPOUND ORGANS.

62. From peculiar combinations of the elementary organs are formed the compound organs.

63. The compound organs are the axis (64) and its appendages (189).

64. The Axis may be compared to the vertebral column of animals.

65. It is formed from an embryo or leaf-bud, by the developement of a root in one direction, and of a stem in the opposite direction.

66. An *embryo* is a young plant, produced by the agency of sexes, and developed within a seed.

67. A *leaf-bud* is a young plant, produced without the agency of sexes, enclosed within rudimentary leaves called scales, and developed on a stem.

68. An embryo propagates the species.

69. Leaf-buds propagate the individual.

70. When the vital action of an embryo or bud is excited, the tissue developes in three directions, upwards, downwards, and horizontally.

71. That part which developes downwards is called the descending axis or root ; that upwards, the ascending axis or stem ; that horizontally, the medullary system ; and the part from which these two axes start is called the crown or collar.

72. This elongation in three directions takes place simultaneously ; hence it follows that all plants must necessarily have an ascending and descending axis, or a stem and root, and a medullary system.

73. The only apparent exceptions to this are the lower tribes of plants, in which the developement seems to be either spherical, filamentous, or horizontal.

III. ROOT.

74. The root is formed by the descending and dividing fibres of the stem.

75. Anatomically it differs from the stem in the absence of normal buds, and of stomates (49), and in Exogens of pith.

76. Although the root has no distinct pith in Exogens, yet it possesses a distinct medullary system.

77. The functions of the root are to fix plants in the earth, and to absorb nutriment from it. As it has to force its way through substances which offer resistance to its passage, it lengthens exclusively by successive additions to the points of its divisions.

78. This absorption takes place almost exclusively by the extremities called *spongelets*, or *spongioles*, which consist of a lax coating of cellular tissue lying upon a concentric layer of woody tissue, in the midst of which is often placed a bundle of ducts (31). Spongioles are not, however, a distinct organ, but are merely the young extremities of roots.

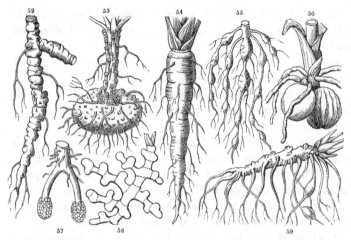

Roots are nodose[52]; placentiform[53]; conical[54]; moniliform[55]; testiculate, or tubercular[56]; coralline[58]; tuberous[59]; and fasciculate, when in clusters as in the Asphodel.

79. Occasionally the epidermis separates from the end of the roots in the form of a cup or cap[57], as in Pandanus and Lycopodium.

80. The power of affording nutriment to the stem and other parts, is not possessed by the root exclusively in consequence of its absorption from the soil. The root is often a reservoir of nutritious matter ready formed, and consisting of starch, as in the Dahlia; mucilage, as in the Orchis; alkaline matter, as in Rhubarb; upon which the young stem feeds, even although the root itself is cut off from communication with any source of supply. Moniliform, tuberous, testiculate, placentiform, conical roots,—in short, all which are unusually thickened,—are intended by nature as reservoirs of food. They must not be confounded with tubers (152), rootstocks (152), or corms (153), all which are forms of stem.

IV. STEM.

81. The stem is produced by the successive developement of leaf-buds (164), which lengthen in opposite directions.

82. If an annular incision be made below a branch of an Exogenous plant (95), the upper lip of the wound heals rapidly, the lower lip not: the part above the incision increases sensibly in diameter, the part below does not.

83. If a ligature be made round the bark, below a branch, the part above the ligature swells, that below it does not swell.

84. Therefore the matter which causes the increase of Exogenous plants in diameter descends.

85. If a growing branch is cut through below a leaf-bud, that branch never increases in diameter between the section and the first bud below it.

86. The diameter of all Exogenous stems increases in each species in proportion to the number of leaf-buds developed.

87. The greater number of leaf-buds above a given part, the greater the diameter of that part ; and *vice versâ*.

88. In the spring the newly forming wood is to be traced in the form of organic fibres descending from the leaf-buds ; that which is most newly formed lying on the outside, and proceeding from the most newly developed buds.

89. Therefore the descending matter, by successive additions of which Exogenous plants increase in diameter, proceeds from the leaf-buds.

90. Their elongation upwards gives rise to new axes, with their appendages; their elongation downwards increases the diameter of that part of the axis which pre-existed, and produces roots.

91. Roots, therefore, in all cases, should consist of extensions of woody tissue ; and this is conformable to observation.

92. Hence, while the stem is formed by the successive evolution of leaf-buds, the root, which is the effect of that evolution, has no leaf-buds.

93. The leaf-buds thus successively developed are firmly held together by the medullary system of the stem, which proceeds from the bark inwards, connecting the circumference with the centre.

94. The stem varies in structure in four principal ways.

95. It is either formed by successive additions to the outside of the wood, when it is called *Exogenous*[60] ; or by successive additions to its centre, when it is called *Endogenous*[62] ; or by the union of the bases of leaves, and by addition to the point of the axis, or by simple elongation or dilatation where no leaves or buds exist ; this is called *Acrogenous*.

96. In what are called *Dictyogens*[61], the stem has the

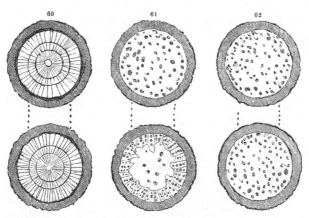

structure of Endogens, the root that of the stem of Exogens nearly ; *Ex.* Smilax.

97. The stem of EXOGENS may be distinguished into the Pith, the Medullary Sheath, the Wood, the Bark, and the Medullary Rays.

98. The PITH consists of cellular tissue, occupying the centre of the stem.

99. It occasionally contains scattered spiral vessels, which appear to originate in the medullary sheath (104), or scattered bundles of vascular and woody tissue, as in Ferula.

100. It is produced by the elongation of the axis upwards.

101. It serves to nourish the young buds until they have acquired the power of procuring nourishment for themselves. For this purpose it is filled with starch, which, in the process of vegetation, becomes converted into mucilage ; and the latter passes out of the pith into the nascent organs.

102. It is always solid when first organized ; but in some cases it separates into regular cavities, as in the Walnut, when it is called *disciform ;* or it tears into irregular spaces, as in Umbelliferous plants.

103. Its office of nourishing the young parts being accomplished, it is of no further importance, and dies.

104. The MEDULLARY SHEATH consists of spiral vessels[63 a].

105. It immediately surrounds the pith, projections of which pass through it into the medullary rays (132)[63 b].

106. It is in direct communication with the leaf-buds and the veins of the leaves.

107. It carries upwards the oxygen liberated by the decomposition of carbonic acid and water, and conducts it into the leaves.

108. The WOOD lies upon the medullary sheath, and consists of concentric layers.

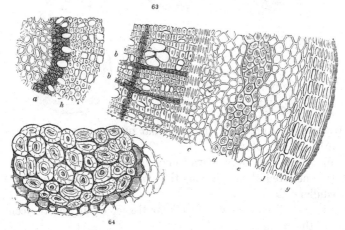

109. It is formed by the successive deposit of organized matter descending from the buds, and by the interposition of the medullary system, here called medullary rays, connecting the pith and the bark[63 b b].

110. The first concentric layer lies immediately upon the medullary sheath and pith, and consists of woody and vasiform tissue[63 h].

111. Each succeeding concentric layer consists of woody and vasiform tissue, which either form themselves into distinct strata, in which case the latter is innermost, or are confounded together.

112. When there is any material difference between the compactness of the tissue of the two sides of a concentric layer, zones are formed in which the woody tissue is outermost; but when the vasiform and woody tissues are equally intermingled, no apparent zones exist.

113. A concentric layer, once formed, never alters in dimensions.

114. Each concentric layer, which is distinctly limited, is usually the produce of one year's growth.

115. Therefore, the age of an Exogenous tree should be

known by the number of concentric circles of the wood.   But
this rule is of uncertain application, owing to numerous dis-
turbing causes, especially in countries in which the period
of rest is less distinctly marked than in the winter of northern
latitudes.

116. The secretions of plants are deposited most abundantly
in the oldest concentric layers ; while those layers which are
most recently formed contain but a slight deposit.

117. When the tissue of the concentric layers is filled with
secretions, it ceases to perform any vital functions.

118. The dead and fully formed central layers are called
the *heart-wood*.

119. The living and incompletely formed external layers
are called the *alburnum*.

120. Upon the outside of the wood lies the BARK, which,
like the wood, consists of concentric layers.

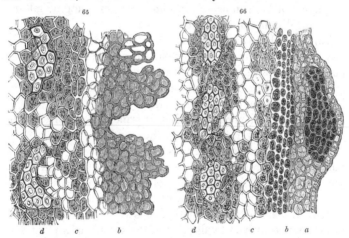

121. It consists of four distinct parts : 1. the *Epidermis* [66 a];
2. the *Epiphlœum* [65 66 b] ;   3. the *Mesophlœum* [65 66 c] ;   and 4,
the *Endophlœum* or *Liber* [65 66 d].

122. Each of these parts increases by successive additions
to its own inside, except the epidermis, which is never re-
newed.

123. The Epiphlœum and Mesophlœum are both formed of
cellular tissue only ; but their cells are placed in different directions with respect to each other.   The former is often
large and soft, and may separate spontaneously from the young

layers forming beneath it, as in Cork, which is the epiphlœum of Quercus Suber.

124. The Endophlœum or Liber consists of cellular tissue resting on the alburnum, of laticiferous tissue (36), and of pleurenchyma (23). The tubes of the latter are often thickened rapidly by a deposit of sedimentary matter; in which case, sections of the tubes present the appearance of concentric circles [64] [65] [66]. Hence arises the toughness of the tubes of pleurenchyma which occur in the liber, and are manufactured into cordage, as in the Lace-bark tree, the Lime-tree, &c.

125. Occasionally the liber is only formed during the first year's growth; after which it is enclosed in wood, and is eventually found near the pith. This has as yet been observed only in the Menispermaceous order.

126. The power of renewing themselves by the production of new matter upon their inner surface, is apparently given to the layers of bark in order to compensate for the gradual and incessant distension of the wood beneath them.

127. As the older parts die, from becoming too small to bear the strain upon them, new parts form, each in its allotted place, and take the station of that which went before it.

128. The secretions of a plant are often deposited in the bark in preference to any other part.

129. Hence chemical or medicinal principles are often to be sought in the bark rather than in the wood.

130. The immediate functions of the bark are to protect the young wood from injury, and to serve as a filter through which the descending elaborated juices of a plant may pass horizontally into the stem, or downwards into the root.

131. It also contains the laticiferous vessels (36), by which the latex is conveyed to all parts of the surface of a plant.

132. The MEDULLARY RAYS or PLATES consist of compressed parallelograms of cellular tissue (*muriform cellular tissue*), belonging to the medullary system.

133. They connect together the tissue of the trunk, maintaining a communication between the centre and the circumference.

134. They act as braces to the woody and vasiform tissue of the wood. They convey secreted matter horizontally from the bark to the heart-wood, and they generate adventitious leaf-buds.

135. *Cambium* is a viscid secretion, which, in the spring, separates the alburnum of an Exogenous plant from the liber. It is free vegetable mucilage, out of which the new elementary organs (8) are constructed, whether in the form of vessels, or woody tissue, or of the cellular tissue of the medullary system, whose office is to extend the medullary plates, and maintain the communication between the bark and central part of a stem.

136. As Exogenous plants increase by annual addition of new matter to their outside, and as their protecting integument or bark is capable of distension in any degree, commensurate with the increase of the wood that forms below it, it follows, taking all circumstances into consideration, that there are no assignable limits to the life of an Exogenous tree.

137. The stem of ENDOGENOUS plants offers no absolute distinction of Pith, Medullary Rays, Wood, and Bark.

138. It is formed by the intermixture of bundles of vascular tissue among a mass of cellular tissue, the whole of which is surrounded by a zone of cellular and woody tissue, inseparable from the stem itself, and therefore not bark.

139. It increases by the successive descent of new bundles of fibro-vascular tissue down into the central cellular tissue, curving outwards as they descend.

140. The vascular bundles of the centre gradually force outwards those which were first formed, the cellular mass augments simultaneously, and in this way the diameter of a stem increases.

141. What appears to be bark in these plants is an external layer of cellular tissue, into which the lower extremities of the arcs of fibro-vascular tissue descend obliquely, losing their vascularity as soon as they reach the cortical integument, or *false-bark*.

142. It is in consequénce of this continuity in an oblique direction of the fibro-vascular bundles and the external cortical integument, that the latter can never, in Endogens, be separated from the wood beneath it.

143. The diameter of the stem of an Endogenous plant is determined by the power its tissue possesses of distending, and by its hardness.

144. When the external tissue has once become indurated, the stem can increase no further in diameter.

145. When the tissue is soft and capable of continual dis-

tension, there is no more certain limit to the life of an
Endogenous than of an Exogenous tree.

146. Generally, the terminal bud only of Endogenous plants
is developed; but very often a considerable number develope;
Ex. Asparagus.

147. When a terminal bud only of an Endogenous plant
developes, the stem is cylindrical; *Ex.* Palms: when several
develope, it becomes conical; *Ex.* Bamboo.

148. In *Acrogens* no other stem is formed than what arises
from the simple union between the bases of the leaves and the
original axis of the bud from which they spring, and which
they carry up along with them.

149. In the order of Ferns the section of a stem indicates
the same structure as that of the numerous petioles (197) out
of which it is constituted.

150. When Acrogens have no proper leaves, they are mere
expansions of cellular matter, sometimes in all directions; *Ex.*
Fungi: sometimes in particular directions; *Ex.* Lichens,
Algæ, &c.

151. The stem of a plant assumes numerous and very different appearances in different plants.

If *above ground* it is rootshaped, or knotted[67]; ascending[68]; creeping[72]; articulated[73]; leafless, succulent, and deformed[69]; or leafy[71].
If it *bears the flowers*, proceeding immediately from the soil or near it, it is a
scape[70].

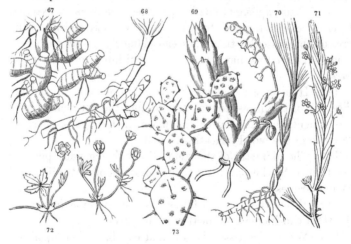

152. It often burrows beneath the earth, when it is vulgarly called a *creeping root*[68]. Sometimes the internodes become much thickened, when what are called *tubers*[77] [78] are formed; or the stem lies prostrate upon the earth, emitting roots from its under side, when it is called a *rhizoma*, or rootstock[67].

153. If it distend underground, without creeping or rooting, but always retaining a round or oval figure, it is called a *corm*[74] [75].

154. All these forms of stem are vulgarly called roots.

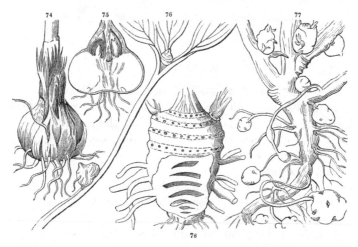

155. No root can have either scales, which are the rudiments of leaves, or nodes, which are the rudiments of buds. A *scaly root* is, therefore, a contradiction in terms.

156. The ascending axis, or stem, has nodes and internodes.

157. *Nodes* are the places where the leaves are expanded and the buds formed.

158. *Internodes* are the spaces between the nodes.

159. Whatever is produced by the evolution of a leaf-bud (164) is a branch.

160. A *spine* is the imperfect evolution of a leaf-bud, and is, therefore, a branch.

161. All processes of the stem which are not the evolutions of leaf-buds, are mere dilatations of the cellular integument of the bark. Such are *prickles* (61).

162. In solid form the stem is extremely variable; the following are common terms relating to it :—

Terete[74] ; half-terete[75] ; compressed[76] ; plano-compressed[76] ; two-edged[77] ; acute-angled[78] ; obtuse-angled[81] ; triangular[83] ; quadrangular[81] ; quinquangular[82] ; octangular[80] ; multangular[80] ; triquetrous[78] ; quadriquetrous[79] ; obscurely triquetrous[85] ; trilateral[86] ; quadrilateral[87] ; quinquelateral[88].

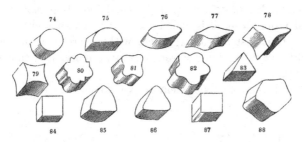

V. LEAF-BUDS.

163. Buds are of two kinds, Leaf-buds and Flower-buds.

164. Leaf-buds (*Bourgeon*, Fr.) consist of rudimentary leaves surrounding a growing vital point, the tissue of which is capable of elongation, upwards in the form of stem, and downwards in the form of root.

165. Flower-buds (*Bouton*, Fr.) consist of rudimentary leaves surrounding a fixed vital point, and assuming, when fully developed, the form of floral envelopes or sexual apparatus.

166. Notwithstanding this difference, a leaf-bud sometimes indicates a tendency to become a flower-bud; and flower-buds frequently assume the characters of leaf-buds ; *Ex.* Monstrous Pears.

167. In appearance a leaf-bud seems[89] to be a collection of scales arranged symmetrically one above the other. These scales are the rudimentary leaves. The centre upon which they are placed is cellular substance coated with a thin stratum of spiral vessels, and these two parts answer to the pith (98) and the medullary sheath (104) in Exogens.

168. By the growth of a leaf-bud a branch is formed ; and the scales gradually change into true leaves as vegetation advances[92].

169. Sometimes they separate spontaneously from the stem

(are deciduous), and are then called *bulbills* or *bulblets*[95]. *Ex.* Lilium bulbiferum.

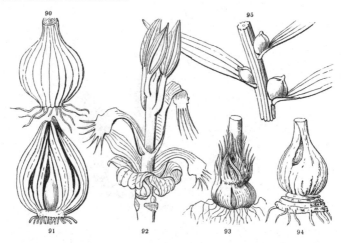

170. Often they are of large size, and are formed underground; they are then called *bulbs* or *scaly bulbs* [90] [91].

171. Although the *corm* (153) is regarded as a kind of underground stem, it may also be considered as a sort of leaf-bud, the centre of which is very large and the scales very thin.

172. In bulbs, young buds or bulbs then called *cloves* (*nuclei*), are often formed in the axils of the scales, as in Garlic; and then gradually destroy the old bulb by feeding upon it. In like manner corms produce other corms at the axils of their scales, and are destroyed by their offspring.

173. Thus in some Gladioli[93] [94], an old corm produces the new one always at its point; the latter is then seated on the remains of its parent, and, being in like manner devoured by its own offspring, becomes the base of the third generation[94]: this process enables such plants by degrees to raise themselves out of the earth in which they were born.

174. In like manner the Crocus[75] produces two or more corms near the apex, and gradually dies as they develope; and the Colchicum bears its mother in the form of a shrivelled spungy lump on one side of its base[96a], while on the opposite side a new bud[96b] is prepared by which the now vigorous parent will hereafter perish.

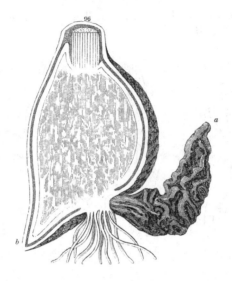

175. Both corms and bulbs are reservoirs of nutriment in either a starchy or mucilaginous condition, or both.

176. Leaf-buds are of two kinds, the regular and the adventitious.

177. *Regular* or normal Leaf-buds are only found in the axils of leaves.

178. They exist in a developed or undeveloped state in the axils of all leaves, and of all modifications of leaves.

179. Therefore they may be expected to appear at the axils of scales of the bud, of stipules (223), of bracts (274), of sepals (335), of petals (336), of stamens (348), and of carpels (406); in all of which situations they are generally undeveloped, for these different organs are all modifications of leaves.

180. They are frequently not called into action, even in the axils of leaves.

181. As regular buds are only found in the axils of leaves, or of their modifications, and as branches are always the developement of buds, it follows, that, whatever may be the arrangement of the leaves, the same will be the disposition of the branches; and *vice versâ*.

182. This corresponding symmetry is, however, continually destroyed by the unequal developement of the buds.

183. Leaf-buds which are formed among the tissue of plants

subsequently to the developement of the stem and leaves, and without reference to the latter, are called latent, adventitious, or abnormal.

184. *Adventitious* Leaf-buds may be produced from any part of the horizontal medullary system, or wherever cellular tissue is present. It has been distinctly proved, that, while roots are prolongations of the vertical or woody system, leaf-buds universally originate in the horizontal or cellular system.

185. They are formed in the root, among the wood, and at the margin or on the surface of leaves.

186. They are constructed anatomically exactly as regular buds, having pith in their centre, surrounded by a medullary sheath of spiral vessels, and coated over by woody tissue and cellular integument.

187. Hence, as adventitious buds, containing spiral vessels, can be produced from parts such as the root or the wood, in which no spiral vessels previously existed, it follows that this form of tissue is either generated spontaneously, or is produced by some other tissue, in a manner unknown to us. It is most probable, that spiral vessels are spontaneous modifications of vesicles of cellular tissue.

188. *Embryo buds* are woody nodules found in the bark of trees, and apparently rudimentary branches formed without leaves, within a space in which they are forcibly pressed upon by the surrounding tissue.

VI. LEAVES.

189. A leaf is an expansion of the bark immediately below the origin of a regular leaf-bud, and is an appendage of the axis (64).

190. Whenever a regular leaf-bud is formed, a leaf, either perfect or rudimentary, is developed also; and *vice versâ*.

191. Leaves are developed alternately[97], one above and opposite the other, around their common axis; but sometimes, in consequence of the internodes being unequally developed, leaves become opposite[98] or verticillate[103]. They are never produced side by side, except by irregular developement.

192. In Exogenous plants, the primordial or seed-leaves (cotyledons) are opposite; hence, in such plants the supposed

non-developement of the axis takes place during the original formation of the embryo.

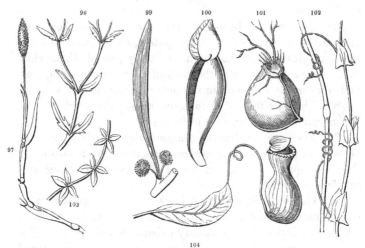

193. There is a constant tendency in opposite or verticillate leaves to become alternate.

194. This law applies equally to the arrangement of all parts that are modifications of leaves.

195. In leaves the developement takes place at their point of junction with the stem; and consequently the tissue at the point of a leaf is the oldest.

196. A leaf consists of a *petiole* or stalk, a *lamina* or blade, and a pair of *stipules*.

197. The PETIOLE is the channel through which the vessels of the leaf are connected with those of the stem; it is formed of one or more bundles of spiral vessels and woody tissue, enclosed in a cellular integument.

198. The spiral vessels of the leaf of Exogenous plants derive their origin from the medullary sheath; those of Endogenous plants from the bundles of fibro-vascular tissue.

199. The cellular integument of the petiole is a continuation of that of the bark.

200. When the petiole is leafy and the lamina is abortive, it is called a *phyllodium*[99].

201. When the petiole becomes dilated and hollowed out at its upper end, the lamina being articulated with and closing

up its orifice, as in Sarracenia[100] and Nepenthes[104], it is called a *pitcher* or *ascidium ;* if it is unclosed, and is a mere sac, as in Utricularia, it is called *ampulla*[101].

202. Sometimes the petiole has no lamina, or is lengthened beyond the lamina, and retains its usual cylindrical or taper figure, but becomes long, and twists spirally ; such a petiole is called a *tendril* (Vrille, *Fr.*)[102].

<blockquote>
The petiole is usually either taper, or channelled ; and it has often a struma[111], (coussinet, Fr.) at either its base or apex, especially in those leaves which are sensitive. In other cases it is inflated[106], sheathing[105], amplexicaul[107], winged[108], auriculate, leafless, jointed[108], spinescent[110], &c.
</blockquote>

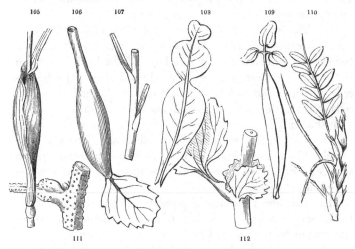

203. The petiole is sometimes *articulated transversely*, as in the Orange.

204. The LAMINA of a leaf is an expansion of the parenchyma of the petiole, and is traversed by veins which are ramifications or extensions of the bundles of vascular tissue of the petiole, or, when there is no petiole, of the stem.

205. Sometimes one, sometimes both the surfaces of a leaf are furnished with stomates.

206. The veins either branch in various directions among the parenchyma, anastomosing and forming a kind of network, or they run parallel to each other, being connected by single transverse unbranched veins.

207. The former is characteristic of Exogenous, the latter of Endogenous plants.

26   STRUCTURAL AND PHYSIOLOGICAL BOTANY.

208. The principal vein of a leaf is a continuation of the petiole, and runs in a direct line from the base to the apex of the lamina; this vein is called the *midrib*. It usually produces other veins from its base or sides, or from both: such veins are called *ribs*, if very strong, and proceeding from the base to the apex; under other circumstances, they are frequently named *nervures*.

209. There are two strata of veins, the one belonging to the upper, and the other to the under surface.

210. The upper stratum conveys the juices from the stem into the lamina, for the purpose of being aërated and elaborated; the under returns them into the bark.

211. The veins are interposed among cellular substance, called *diachyma, diploë*, or *mesophyllum*; which is often stratified differently below the two surfaces of the leaf; the upper stratum being more compact than the lower, and having its cells perpendicular to the plane of the leaf: in such cases, the cells of the lower stratum are commonly more or less parallel with the under surface.

212. The lamina is variously divided and formed; it is usually thin and membranous, with a distinct upper and under surface; but sometimes becomes succulent, when the surfaces are often not distinguishable.

213. The upper surface is presented to the sky, the lower to the earth; this position is rarely departed from in nature, and cannot be altered artificially, except by violence.

214. A leaf is *simple* when its lamina is undivided, or when, if it is separated into several divisions, those divisions do not reach the midrib; *Ex.* Lime-tree, Oak.

215. The form of the simple leaf is extremely variable, and the terms employed to denote the variations are numerous in proportion.

216. Some leaves have the margin so continuous, that the outline is scarcely interrupted, except by small toothings. Of such leaves the following are among the more common forms:

Orbicular[120]; ovate[121]; lanceolate[122]; oval[118]; oblong[117]; roundish oblong[113]; peltate[124]; cordate[119]; cordate ovate[114]; cordate acuminate[119]; reniform[123]; oblique[115]; auriculate[139].

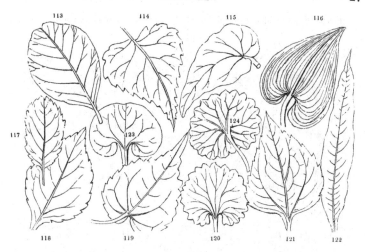

217. In other leaves the margin is produced here and there into manifest angles ; in which cases the following terms are commonly in use :

Sagittate or arrow-headed[126] ; cuneate or wedge-shaped[127] ; hastate[130] ; angular[131] ; triangular[128].

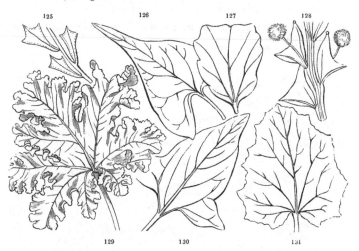

218. In other cases the margin is repeatedly interrupted in a definite manner along its whole course ; and then such terms as the following are employed :

Palmate[138] ; seven-lobed[134] ; pinnatifid[133] ; sinuated[132] ; panduriform[135].

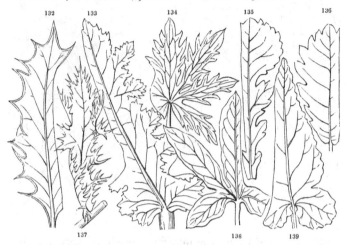

219. A leaf is compound when the divisions pass down to the midrib so as to subdivide the leaf into smaller and distinct leaves, or leaflets (*foliola*).

220. When leaves are compound, their mode of division is expressed by such terms as the following :

Ternate[144] ; biternate[146] or triternate ; digitate[140] ; pedate[142] ; pinnate[145] ; interruptedly pinnate[147] ; lyrate[143] ; bipinnate[150] ; decompound or tripinnate[141] ; bijugate[148] ; conjugato-pinnate[149].

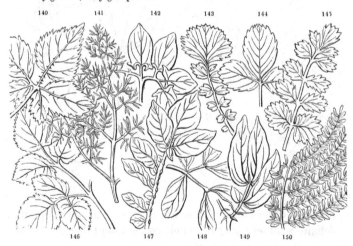

221. In speaking of the margin, we say that it is

Entire[126]; serrate[114]; biserrate[137]; dentate[139]; duplicato-dentate[137]; tri-dentate[125]; crisp or curled[129]; crenate[120].

222. The point of the leaves gives rise to other terms, such as the following:

Acute[118]; obtuse[139]; retuse[113]; emarginate[113]; acuminate[119]; mucronate[132]; truncate[129].

223. STIPULES are attached to each side of the base of the petiole. They have, if leafy, veins, the anatomical structure of which is the same as that of the veins of the leaves.

224. Sometimes only one stipule is formed, the other being constantly abortive, as in Azara.

225. Stipules are sometimes transformed into leaves: they sometimes have buds in their axils; and may be, therefore, considered rudimentary leaves.

226. Whatever arises from the base of a petiole, or of a leaf, if sessile, occupying the same place, and attached to each side, is considered a stipule. The appearance of this organ is so extremely variable, some being large and leaflike, others being mere rudiments of scales, that botanists are obliged to define it by its position, and not by its organization.

227. The stipules must not be confounded with cellular marginal appendages of the petiole, as in Apocynaceæ.

228. Stipules, the margins of which cohere in such a way that they form a membranous tube sheathing the stem, are called *ochreæ; Ex.* Rhubarb.

229. All leaves are originally continuous with the stem; as they grow, an interruption of their tissue at their junction with the stem takes place, by which a more or less complete articulation is formed sooner or later.

230. The articulation between a leaf and stem being completed, the tissue of the former becomes gradually incrusted by the foreign matter deposited by the sap in the process of secretion and digestion, and at last is incapable of further action, when it dies. When the stem continues to increase in diameter, as a dead leaf will not increase with it, the latter is eventually thrown off; this is the fall of the leaf. But in some Endogens the articulation is so slight, and the stem increases so little in diameter, that the leaf is never thrown off, but simply withers and decays.

231. All leaves ultimately fall off; evergreen leaves later than others.

232. The mode in which leaves are arranged within their bud is called *vernation,* or *gemmation.*

233. Leaves have, under particular circumstances, the power of producing leaf-buds from their margin (185); *Ex.* Bryophyllum, Malaxis paludosa, and proliferous Ferns.

VII. FOOD AND SECRETIONS.

234. Plants are nourished by the absorption of food from the air and earth, in consequence of which they grow, and produce their peculiar secretions.

235. The growth of plants is very rapid; that of the leaves is such that they often acquire six or seven times their original weight per hour.

236. The food of plants always consists of carbonic acid, nitrogen, and water, and also of various mineral matters, chiefly alkaline, the nature of which varies according to species.

237. Roots have the power of absorbing most substances in a fluid or gaseous form, even although their extremities are unbroken.

> It appears probable that when plants are incapable of imbibing certain substances, such as strontian, there is no isomorphism between their ordinary mineral constituents and those they reject. Thus, lime and magnesia, which plants will indifferently absorb, are isomorphous ; but between them and strontian, which they will not absorb, no isomorphism exists.—*Daubeny.*

238. Carbon is obtained by plants in the form of carbonic acid, derived from the atmosphere, or generated in soil by the decay of vegetable matter.

239. Hydrogen is obtained principally by the decomposition of water, and is assimilated along with carbonic acid, while the oxygen of the water is liberated.

240. Nitrogen can only be obtained by plants in the form of ammonia. The nitrogen of the atmosphere cannot be the source of supply, because it cannot be made to enter into combination with any element except oxygen, even by the employment of the most powerful chemical means.

241. Ammonia exists in every part of plants, in the roots, in the stem, and in all blossoms and fruits in an unripe condition.

It is supplied by rain-water, which carries it down from the air, in which it is suspended, in consequence of the putrefaction of animal and vegetable matters. This ammonia affords all vegetables, without exception, the nitrogen which enters into the composition of their constituent substances.

242. A certain portion of the ammonia which falls with rain evaporates again with the water; but another portion is taken up by the roots of plants, and, entering into new combinations, produces albumen, gluten, and a number of other compounds, containing nitrogen.

243. But it is not so much the *quantity* of ammonia that is important to plants, as the *form* in which it is presented to them. When in a volatile state, it is in great measure lost before it can be imbibed. When fixed, in the state of salts, its volatility is overcome, and not the smallest portion of the ammonia is lost to the plants, for it is all dissolved by water and imbibed by the roots.

244. But carbonic acid, water, and ammonia, are not the only elements necessary for the support of vegetables. Certain inorganic constituents are also essential.

245. Phosphate of magnesia in combination with ammonia is an invariable constituent of the seeds of all kinds of grasses. The acids found in the different families of plants are of various kinds. It cannot be supposed that their presence and peculiarity are the result of accident. If these acids are constantly present and necessary to life, it is equally certain that some alkaline base is also indispensable, in order to enter into combination with the acids, which are always found in the state of salts.

246. If a plant does not produce more of its peculiar acids than it requires for its own existence, a plant must contain an invariable quantity of alkaline bases, wherewith the vegetable acids may form salts.

247. The proportion of alkaline bases in a plant is indicated by the quantity of ashes they yield. The quantity of ashes obtained from the same quantity of vegetable matter varies constantly in different species. Therefore the proportion of alkaline bases varies in different species, and consequently different species demand a different amount of alkaline food in the soil.

248. The perfect developement of a plant is therefore dependent on the presence of alkalies or alkaline matter; for when these substances are totally wanting, growth will be arrested; and when they are deficient, it must be impeded in proportion.

249. But other substances besides alkalies are required to sustain the life of plants. Phosphoric acid has been found in the ashes of all plants hitherto examined; and common salt, sulphate of potash, nitre, salts of iron and copper, chloride of potassium, and other matters, may be regarded as necessary constituents of several plants.

250. Therefore it is indispensable that every plant should find in the soil it is cultivated in those inorganic constituents which nature has rendered necessary to it, just as it is necessary for animals that they should find in their food the phosphates of lime and magnesia, which harden their bones.

251. As soon as food is absorbed, it begins to ascend into the stem, or to diffuse itself through the system, and receives the name of sap.

252. In the course of the sap upwards, the water and carbonic acid are partially decomposed and their elements are deposited along with nitrogen in the interior of the tissue, forming a layer over the interior of every cell and vessel, which thus become in part solidified.

253. As soon as the sap reaches the leaves or the surface of the bark, green matter, or occasionally some other colour, is formed, provided the part is exposed to light.

254. This appears to arise chiefly from the decomposition of carbonic acid, ammonia, and water, when the carbon, nitrogen, and hydrogen are fixed by the plant, and the oxygen restored to the atmosphere. Such action is called the *assimilating power* of plants.

> Plants are, therefore, the great purifiers of the atmosphere, consuming the products of animal respiration and of all organic putrefaction, and converting them again into matter suited to the wants of man.

255. In the absence of light, plants re-absorb oxygen from the atmosphere, and re-combine it with the matter they contain, to be again liberated at the return of light.

256. They also, at all times, especially at night, part with carbonic acid in small quantities.

It has, however, been proved experimentally that they purify the air much more by their assimilating (254) action, than they vitiate it by their respiration.

257. No plants can long exist in which this alternate action is prevented, unless, perhaps, Fungi and brown parasites.

258. The amount of assimilation is determined by the degree of light to which a plant is exposed. It is light alone that causes, in conjunction with vital forces, the decomposition of the matters contained in living plants.

259. Hence, if a plant is compelled to grow in darkness, no assimilation takes place of the food that the roots receive; oxygen accumulates; its natural proportion to other elements is disarranged; and a destruction of the tissue takes place.

260. In order to avoid this, plants will always lengthen themselves in the direction in which the smallest ray of light approaches them, as is the case of seed which shoot from darkness into light. If this is impossible, they become blanched or *etiolated*, and then die.

261. From the continued assimilation of the elementary constituents of plants, new products result, and serve for the formation of woody fibre, and all solid matters of a similar composition. The leaves produce sugar, starch, and acids, which were previously formed by roots, when necessary for the developement of the stem, buds, leaves, and branches.

Some phyto-chemists believe that during the chemical transformations that result in plants from the separation and re-combination of their elements, two compounds are necessarily formed, one of which remains as a component part, while the other is separated by the roots, in the form of excrementitious matter. But the experiments upon which this supposition is founded are not considered conclusive; and great doubt is entertained whether plants have really the power of rejecting excrementitious matter by their roots. It appears more probable that the necessary separation of effete matter takes place by the hairs and glands that clothe the surface of plants, or by a fluid secretion from their whole surface.

262. Sap (251) is put in motion by the newly developing leaf-buds, which, by constantly consuming the sap that is near them, attract it upwards from the roots as it is required. Therefore, the movement of the sap is the effect, and not the cause, of the growth of plants. It depends upon vital irritability, and is independent of mechanical causes.

263. This *irritability* is indicated not only by the motion of the sap, but by several other phenomena of vegetation; such as,

The elasticity with which the stamens sometimes spring up when touched, and the sudden collapse of many leaves when stimulated; the apparently spon-

taneous oscillation of the labellum of some Orchidaceous plants ; the expansion
of flowers and leaves under the stimulus of light, and the collapse of them
when light is withdrawn (this phenomenon in leaves is called the *sleep of
plants*) ; and by the effects of mineral and vegetable poisons being the same
upon plants as upon animals.   Mineral poisons kill by inflammation and cor-
rosion ; vegetable poisons by the destruction of irritability.

264. After the sap has been distributed through the veins
of the leaves, it becomes exposed to the influence of air and
light, and undergoes peculiar chemical changes.   In this state
it is called the *proper juice*.

265. When the proper juice has been once formed, it flows
back, and descends towards the roots, passing off horizontally
into the centre of the stem.

266. Hence the great importance of leaves to plants, and
the necessity of exposing them to the full influence of light and
air, for the purpose of securing a due execution of their natural functions.   Hence also the impropriety of mutilating
plants by the destruction of their leaves.

267. In Exogenous plants (95), the upward course of the
fluids is through the young wood; their downward passage
through the bark, towards, or into the root; and their horizontal diffusion takes place by the medullary rays.

268. Hence the peculiar principles of such plants are, in
trees and shrubs, to be sought either in the bark or the heart-
wood (118), not in the alburnum (119).   But in plants whose
stems are annually destroyed while the roots are perennial, the latter are the chief reservoir of secretions; and in
annuals, whose root and stem both perish, the secretions are
dispersed equally through the stem and root.

269. As they are the result of the growth of a plant, they
will be found more abundantly in annual plants at the end
than at the commencement of their growth.

270. In Endogenous plants (95) it is probable that the upward course of the fluids is through the bundles of vascular
and woody tissue, and that the downward and horizontal passage takes place through the cellular tissue.

271. The precise direction of the sap in Acrogens (95) is
unknown.

VIII. FLOWER-BUD.

272. The FLOWER-BUD consists of a fixed point, surrounded by imbricated, rudimentary, or metamorphosed leaves,

the external or inferior of which are usually alternate, and the internal or superior always verticillate, or opposite; the latter are called *floral envelopes* and *sexes*.

273. As every flower-bud proceeds from the axil of a leaf, either fully developed or rudimentary, it therefore occupies exactly the same position with respect to the leaf as a leaf-bud.

274. The leaf from the axil of which a flower-bud arises, is called *bract* or *flower-leaf*; and all rudimentary leaves, of what size or colour soever, which appear on the peduncle (284) between the floral leaf and the calyx (325), are called *bracteolæ* or *bractlets*.

275. But, in common language, botanists constantly confound these two kinds, which are, nevertheless, essentially distinct.

276. Although the buds in the axils of bracts are often not developed, yet they have the same power of developement as those in the axils of leaves; they are generally flower-buds, very rarely leaf-buds.

277. When a single bract is rolled together, highly developed, and coloured, and is placed at the base of that form of inflorescence called a spadix (304), it is named *spathe*; *Ex*. Arum.

278. When several bracts are verticillate or densely imbricated around the base of the forms of inflorescence, called the umbel or capitulum (306), they receive the name of *involucre*; *Ex*. Carrot, Daisy.

279. When the bracts of an involucre form a single whorl, and cohere by their margins, it is impossible to distinguish them from the calyx by any other mark than by their position, and by their usually surrounding more flowers than one.

280. The minute or colourless bracts at the base of the florets of a capitulum (306) are called *paleæ*.

281. Small imbricated bracts are often called *scales*.

282. Bracts, when placed immediately below the sexes, as in apetalous flowers, are only distinguished from the calyx by being alternate with each other, and not verticillate; hence the *glumes* and *paleæ* of grasses are bracts and not calyx.

283. The axis of the flower-bud in its natural state does

not lengthen beyond those upper series of metamorphosed leaves which constitute the sexes.

284. The lengthened part of the axis, from the point of its connection with the stem, as far as the floral envelopes, is called the *peduncle*.

285. When several peduncles spring from the axis at short distances from each other, the axis receives the name of *rachis*, and the peduncles themselves are called *pedicels*.

286. There is never more than one flower to each peduncle, strictly speaking; therefore, when we speak of a two-flowered peduncle, we only mean that two flowers, each having its peculiar pedicel, terminate the axis, which is then considered a peduncle common to each pedicel.

287. Every flower, with its peduncle and bractlets, being the developement of a flower-bud, and flower-buds being altogether analogous to leaf-buds, it follows, as a corollary, that every flower, with its peduncle and bractlets, is a metamorphosed branch.

288. And further, the flowers being abortive branches, whatever are the laws of the arrangement of branches with respect to each other, the same will be the laws of the arrangement of flowers with respect to each other.

289. Flower-buds, however, being much less subject to abortion than leaf-buds, flowers are more symmetrically disposed than branches, and appear to possess their own peculiar order of developement.

290. As flower-buds can only develope from the axil of a bract, it follows, that while a pedicel without bracts can never accidentally produce other flowers, any one-flowered pedicel, on which bracts are present, can, and frequently does, bear several flowers.

291. In consequence of a flower and its peduncle being a branch in a particular state, the rudimentary or metamorphosed leaves which constitute bracts, floral envelopes, and sexes, are subject to exactly the same laws of arrangement as regularly formed leaves.

292. The manner in which the floral organs, especially the calyx and corolla, are arranged before expansion takes place, is called the *æstivation* or *præfloration*.

The following are the principal kinds of æstivation :—valvate[151] ; valvate and involute[156] ; imbricate[157] ; alternate[159] [160] ; convolute[152] ; induplicate[155] ; plicative[153] ; quincuncial[157] [158] ; supervolutive[154] ; vexillary[161].

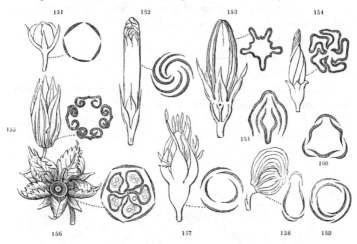

293. The modes in which the flower-buds are arranged are called *forms of inflorescence ;* and the order in which they unfold is called the *order of expansion.*

IX. INFLORESCENCE.

294. Inflorescence is the ramification of that part of the plant intended for reproduction by seed.

295. The greater developement of some forms of inflorescence than of others, is owing to the greater power one plant possesses than another of developing buds, latent in the axils of the bracts.

296. In consequence of flower-buds obeying the laws which regulate leaf-buds, all forms of inflorescence must, of necessity, be axillary to a leaf of some kind.

297. Those forms which are called *opposite the leaves, extra-axillary, petiolar* or *epiphyllous*, and even the *terminal* itself, are mere modifications of the axillary.

298. The kinds of inflorescence which botanists more particularly distinguish are the following :

299. When no elongation of the general axis of a plant takes place beyond the developement of a flower-bud, the flower becomes what is called *terminal* and *solitary ; Ex.* Pæony.

300. When a single flower-bud unfolds in the axil of a leaf, and the general axis continues to lengthen, and the leaf undergoes no sensible diminution of size, the flower which is developed is said to be *solitary* and *axillary*.

301. If all the buds of a newly formed elongated branch develope as flower-buds, and at the same time produce peduncles, a *raceme* is formed[163].

302. If buds, under the same circumstances, develope without forming peduncles, a *spike* is produced[162].

303. Hence the only difference between a spike and raceme is, that in the former the flowers are sessile, and in the latter stalked.

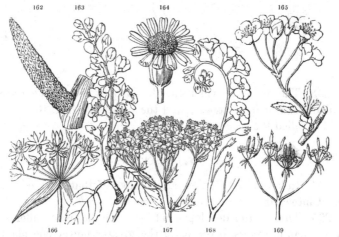

304. A *spadix* differs from a spike in nothing more than in the flowers being packed close together upon a succulent axis, which is enveloped in a spathe (277).

305. An *amentum* is a spike the bracts of which are all of equal size, and closely imbricated, and which is articulated with the stem.

306. When a bud produces flower-buds, with little elongation of its own axis, either a *capitulum*[170 172], or an *umbel*[169], is produced.

307. The capitulum bears the same relation to the umbel as the spike to the raceme; that is to say, these two forms differ in the flower-buds of the capitulum being sessile, and of the umbel having pedicels.

308. The dilated depressed axis of the capitulum is called the *receptacle*.

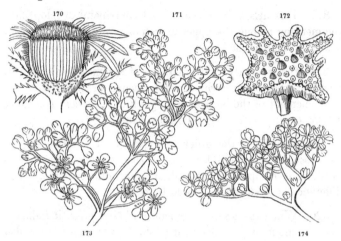

309. A raceme, or panicle, the lowest flowers of which have long pedicels, and the uppermost short ones, is a *corymb* [165] [167].

310. A *panicle* is a raceme, the flower-buds of which have, in elongating, developed other flower-buds [173].

311. A panicle, the middle branches of which are longer than those of the base or apex, is called a *thyrsus*.

312. A panicle, the elongation of all the ramifications of which is arrested, so that it assumes the appearance of an umbel, is called a *cyme* [174].

313. In all modes of inflorescence which proceed from the buds of a single branch, the axis of which is either elongated or not, the flowers expand first at the base of the inflorescence, and last at the summit. This kind of expansion is called *centripetal*.

314. When the uppermost or central flowers open first, and those at the base or the circumference last, the expansion is called *centrifugal*.

315. The centripetal order of expansion always indicates that the inflorescence proceeds from the developement of the buds of a single branch.

316. When inflorescence is the result of the developement of several branches, each particular branch follows the centri-

petal law of expansion, but the whole mass of inflorescence the centrifugal.

317. This arises from the partial centripetal developement commencing among the upper extremities of the inflorescence, instead of among the lower.

318. Consequently, this difference of expansion will indicate whether a particular form of inflorescence proceeds from the developement of the buds of a single branch, when it is called *simple*, or not, when it is called *compound*.

319. Whenever the order of expansion is centripetal, the inflorescence is to be understood as simple; when it is centrifugal, it is compound, although in appearance simple. This difference is often of great importance.

320. When the order of expansion is irregular, it indicates that the mode of developement of the flowers is irregular also, either on account of abortion or other causes.

321. Sometimes all the flowers of the inflorescence are abortive, and the ramifications, or the axis itself, assume a twisted or spiral direction; when this happens, a *tendril* is formed; *Ex*. the Vine.

X. FLORAL ENVELOPES.

322. The Floral Envelopes are the parts which immediately surround the sexual organs.

323. They are formed of one or more whorls of bracts, and are therefore modified leaves (274).

324. In anatomical structure they do not essentially differ from the leaves, farther than is necessarily consequent upon the peculiar modifications of size or developement to which they are subject.

325. When the floral envelopes consist of but one whorl of leaves, they are called *calyx*.

326. When two or more whorls are developed, the outer is called calyx, the inner *corolla*.

327. There is no other essential difference between the calyx and corolla. Therefore, when a plant has but one floral envelope, that one is calyx, whatever may be its colour or degree of developement.

328. It is necessary, however, to be aware, that sometimes the calyx is reduced to a mere rim, either in consequence of lateral compression, as in the *pappus* (*aigrette*, Fr.) of many Compositæ, or from other unknown causes, as in some Acanthaceæ.

329. If the floral envelopes are of such a nature that it is not obvious whether they consist of both calyx and corolla, or of calyx only, they receive the name of *perianthium* or *perigonium*.

330. Plants have frequently no floral envelopes; in that case flowers are said to be *naked* or *achlamydeous*.

331. When the floral envelopes are deciduous, they fall from the peduncle, as leaves from a branch, by means of an articulation; if they are persistent, it is because no articulation exists.

332. When the margins of floral envelopes are united, the part where the union has taken place is called the *tube*, and that where they are separate is named the *limb*. It frequently happens that in the calyx an articulation forms between the limb and the tube.

333. Botanists generally consider that the tube of the calyx is invariably formed by the union of the margins of the sepals. It is, however, probable that it is in some cases a mere dilatation and expansion of the pedicel itself, as in Eschscholtzia.

334. When the calyx and corolla are readily distinguishable from each other, they exhibit the following peculiarities:

335. The *calyx* consists of two or more divisions, usually green, called *sepals*, which are either distinct, when a calyx is said to be *polysepalous*; or which unite by their margins in a greater or less degree, when it is called *monosepalous, gamosepalous*, or *monophyllous*.

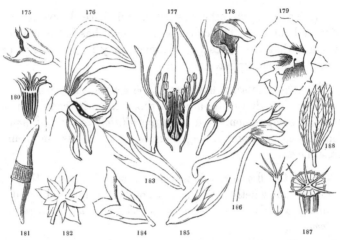

The calyx may be superior[178], or inferior[177]; galeate[176]; calyptrate[181]; double[182]; calcarate[186]; coroniform[187]; vesicate[188]; dilated[179]; spiny[183]; oblique[175][185]; ringent[184].

336. The *corolla* consists of two or more divisions, called *petals*, usually of some bright colour, different from that of the sepals, than which they are frequently more developed. When the petals are distinct, a corolla is said to be *polypetalous;* when they are united by their margins, it is called *gamopetalous* or *monopetalous*.

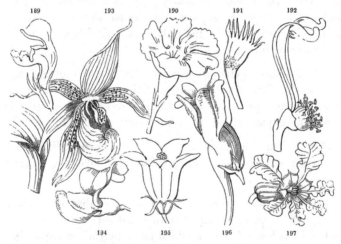

The corolla may be labiate[189]; calceolate[193]; ringent[196]; papilionaceous[191]; campanulate[195]; funnel-shaped[190]; crisp[197].

337. If the union of the petals or sepals takes place in one or two parcels, the corolla or calyx are said to be *one or two-lipped*. These lips are always anterior and posterior with respect to the axis of inflorescence, and never right and left.

338. If the sepals or petals are of unequal size, or unite in unequal degrees, the calyx or corolla is said to be *irregular*[189].

339. If the sepals and petals are unequal in number, or no multiple of each other, or if the stamens are neither equal to them in number, nor any power of them, a flower is said to be *unsymmetrical*.

340. When the petals are so arranged, that of five the uppermost is dilated, the two lateral ones contracted and parallel with each other, and the two lower also contracted, parallel with each other, and coherent by their anterior margins, a flower is said to be *papilionaceous*[194].

341. When a petal tapers conspicuously towards the base, it is said to be *unguiculate*[191]; its lower part is called the *unguis*, its upper the *limb*. The former is analogous to the petiole, the latter to the lamina of a leaf.

342. The petals always alternate with the sepals, a necessary consequence of their following the laws of developement of leaves.

343. If at any time the petals arise from before the sepals, such a circumstance is due to the abortion of one whorl of petals between the sepals and those petals which are actually developed.

344. As petals always alternate with sepals, the number of each row of either will always be exactly the same. All deviations from this law are either apparent only, in consequence of partial cohesions, or, if real, are due to partial abortions.

345. Whatever intervenes between the bracts and the stamens belongs to the floral envelopes, and is either calyx or corolla; of which nature are many of the organs vulgarly called *nectaries*.

Of this nature are the horn-like bodies found beneath the upper galeate sepal of Aconitum[192], the cup of Narcissus[200], a part of the coronal appendages or coronet of Stapelia[199 203].

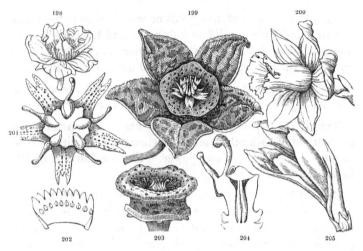

346. But it is to be observed, that as there are no exact limits between the corolla and the stamens (348), such bodies as have been just described are often of an *indifferent* nature, and may be referred with equal justice to petals passing into stamens, and to stamens passing into petals.

This is particularly the case with the fringes of Parnassia[198], some parts of the coronet of Stapelia[203], the long rays of the Passion-flower.

347. If, however, anomalous bodies at this part of the vegetable system can be shown to belong to any whorl or series of which a part is certainly petals or stamens, such anomalous bodies are to be regarded as belonging to the organ in whose series they are placed.

Thus in Aconitum[192], the horn-like processes belong to the series of the corolla, and are therefore petals ; in the Mahogany[202], and in the Canna[205], they evidently appertain to the Andrœceum (348), and are therefore stamens. This settles the true nature of what has been called the nectary[206], in Orchidaceous plants, now termed the lip, or labellum, which, forming a part of the second series of floral envelopes, is therefore universally recognised as a petal, notwithstanding its singular form.

XI. MALE ORGANS.

348. The whorl of organs immediately within the petals, is composed of bodies called *stamens*, which are considered the male apparatus of plants, and constitute the *Andrœceum*.

349. They consist of a bundle of spiral vessels surrounded by cellular tissue, called the *filament*, terminated by a peculiar arrangement of the cellular tissue, in a case, finally opening and discharging its contents, called the *anther*.

350. There are many instances in which no limits can be traced between the petals and stamens; *Ex.* Nymphæa.

351. In such cases it is found that the limb (341) of the petal contracts, and becomes an anther, while the unguis assumes the state of a filament.

352. Now as there are no limits between the petals and sepals (327), nor between the sepals and bracts (323), nor between the bracts and leaves (274), it follows that the stamens are also a modification of leaves.

353. And as the limb of a petal is analogous to the lamina, and the unguis (341) to the petiole of a leaf, it also follows that the anther is a modification of the lamina, and the filament of the petiole.

354. The stamens follow the same laws of successive developement as leaves; and, consequently, if their arrangement be normal, they will be either equal in number to the petals, and alternate with them, or, if more numerous, some regular multiple of the petals.

355. If they are twice the number of petals, two whorls are considered to be developed; and so on.

356. If they are equal in number to the petals, and opposite them, it is to be understood that the innermost only of two whorls is developed, the outermost being abortive.

357. All deviations from these laws are owing to the abortion of some part of the stamens; *Ex.* Lamium, Hippuris.

358. When the stamens do not contract any union with the sides of the calyx, they are *hypogynous*[218]; *Ex.* Ranunculus.

359. When they contract adhesion with the sides of the calyx, they become *perigynous*; *Ex.* Rose[177].

360. If they are united both with the surface of the calyx and of the ovary, they are *epigynous ; Ex.* Umbelliferæ.

361. When two are long and two are short[217], they are called *didynamous ;* and if out of six two opposite ones are shorter than the other four, they are *tetradynamous.*

362. The *filaments* (349) are either distinct or united by their margins. If they are united in one tube, they are called *monadelphous*[216] *; Ex.* Malva: if in two parcels, *diadelphous*[219] *; Ex.* Pea: if in several, *polyadelphous*[218] *; Ex.* Hypericum.

363. When they are united in a solid body, along with the style, they form what is called a *column,* and are said to be *gynandrous*[206].

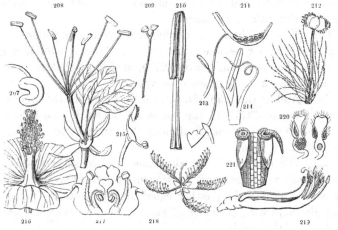

Filaments are sometimes apparently forked[209], in consequence of the separation of the connective (366), into two arms ; strumose, when a tubercle forms upon their face[220] ; stupose, if covered with long hairs[212] ; and toothed[214], if their margin is lengthened on either side beyond the attachment of the anther[214].

364. The filament is not essential to a stamen, and is often absent.

365. The *anther*[210] is the limb of the stamen, forming within its substance, and finally emitting a matter called *pollen.*

366. The two sides of the anther are called its *lobes ;* and the solid substance which connects them, and which is in fact a continuation of the filament, as the midrib of a leaf is of the petiole, is named the *connective.*

367. The connective is usually simple and uninterrupted ; but it is sometimes lengthened into two arms[209], or is articu-

lated with the filament, across which it is placed, and on which it swings. In the latter case it either bears an anther-lobe on both arms[215], or only on one[213]; *Ex.* Salvia.

368. The cavities of the anther containing the pollen are the *cells*, and the place by which the pollen is emitted is the point or line of *dehiscence;* the membranous sides of the anther are named the *valves.*

369. Dehiscence usually takes place along a line, which may be considered to indicate the margin of the limb out of which the anther is formed ; *Ex.* Rose.

370. Sometimes a portion only of this line opens, and then the anther is said to dehisce by *pores; Ex.* Azalea.

371. If the line of dehiscence occupies both margins of the connective, and not the centre of the lobes, the anther opens by one valve instead of two, which is then hinged by its upper edge; *Ex.* Berberry.

372. The cells of the anther are usually two in number: sometimes they are four[239]; *Ex.* Tetratheca : rarely one; *Ex.* Epacris: and still more rarely several; *Ex.* Viscum [223].

373. The number of cells appears to be determined by no certain rule.

374. Sometimes the cells are folded down upon themselves and become sinuous[237]; in other cases they are prolonged into bristles[227] [240], or tubes[224], or even into a spur[211]; *Ex.* Mela-stomaceæ.

375. Although in most cases the line of dehiscence is parallel with the anther-lobes, it is occasionally transverse[238]. In Laurus the transverse and hinged (371) dehiscence being combined[226], the face of the anther breaks up into four hinged lobes.

376. It may be conjectured that the transverse dehiscence of an anther is analogous to the transverse articulation of petioles (203).

377. The anthers frequently grow together by their mar-gin; *Ex.* Compositæ. Such anthers are called *syngenesious.*

378. The *Pollen* is formed by a peculiar modification of the cellules of the parenchyma of the anther.

379. It consists of hollow cases, of extreme smallness, con-taining a fluid in which float grains of starch and drops of oil.

380. It is furnished with apertures[229], through which its

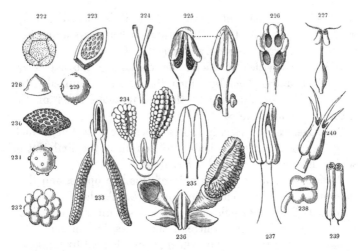

lining is protruded in the form of a delicate tube, where the pollen comes in contact with the stigma.

381. The shape of pollen grains is very variable; the spherical[231], the triangular[228], the polygonal[222], the oblong[230], are common forms.

382. Its surface is either smooth or studded with little points[231].

383. The pollen grains are usually distinct from each other, but in some cases they cohere in, definite numbers; *Ex.* Acacia[232]: or in irregular masses; *Ex.* Orchidaceæ[234]: or are enclosed within a bag, which seems to be the lining of the anther (Endothecium); *Ex.* Asclepiadaceæ[233] [236].

384. In cases where the pollen grains cohere in masses, or are enclosed within bags, they are connected with a cartilaginous or elastic process, called the *caudicle*[234], which adheres to a gland[234] belonging to the stigma.

385. The function of the pollen is to vivify the *ovules* (396).

XII. DISK.

386. Whatever intervenes between the stamens and the pistil receives the general name of disk.

387. It usually consists of an annular elevation, encompassing the base of the ovary, when it is sometimes called the *cup*; *Ex.* Pæony.

388. Or it appears in the form of a glandular lining of the

tube of the calyx ; *Ex.* Rose : or of tooth-like, hypogynous (358) processes ; *Ex.* Gesnera, Cruciferæ.

389. When a fleshy substance occupies the centre of a flower, and bears a single row of carpels, it is called the *gynobase ; Ex.* Lamium, Ochna, Geranium, &c. If this substance bears a greater number of carpels than can be arranged in one row, it is called the *torus* or *receptacle ; Ex.* Strawberry, Nelumbium.

390. It is certain that the disk is a non-developement of an inner row or rows of stamens, as is proved by the Moutan Pæony.

391. The receptacle or torus is the growing point (164) of the flower-bud in a state of enlargement.

392. The disk is one of the parts which Linnæan botanists call *nectary*.

XIII. FEMALE ORGANS.

393. The organ which occupies the centre of a flower, within the stamens and disk, if the latter be present, is called the *pistil*.

394. It is the female apparatus of flowering plants, or the *gynœceum* [241].

395. It is distinguished into three parts; viz. the *ovary*, the *style*, and the *stigma*.

396. The Ovary is a hollow case, enclosing *ovules* (445). It contains one or more cavities, called *cells*.

397. The Stigma is the upper extremity of the pistil.

398. The Style is the part that connects the ovary and stigma.

399. The style is frequently absent, and is no more essential to a pistil than a petiole to a leaf, or a filament to an anther.

400. Sometimes the style is thin, flat, and membranous, and assumes the form of a petal, as in Iris.

401. The style is either articulated with the ovary, or continuous with it. It usually proceeds directly from the apex of the ovary ; but in some cases arises from the side, or even the base of that organ ; *Ex.* Alchemilla, Chrysobalanaceæ.

402. Nothing is, properly speaking, stigma, except the secreting surface of the style. Nevertheless, the name is often inaccurately applied to mere divisions of the style, as in Labiatæ; or to the hairy surface of undivided styles, as in Lathyrus.

403. Sometimes the stigmas grow to the face of the anthers, which form themselves into a solid mass; Ex. Asclepias[204]. In this case the styles remain separate.

404. The pistil is either the modification of a single leaf, or of one or more whorls of modified leaves.

405. Such modified leaves are called *carpels*.

406. A Carpel is formed by a folded leaf, the upper surface of which is turned inwards, the lower outwards; and within which are developed one or a greater number of buds, which are the *ovules*.

407. When the carpels are stalked, they are said to be seated upon a *thecaphore*, or *gynophore*; Ex. Cleome, Passiflora. Their stalk is analogous to the petiole of a leaf.

408. When the carpels are all distinct, or are separable with facility, they are *apocarpous*; when they all grow into a solid body, which cannot be separated into its constituent parts, they are *syncarpous*.

409. The ovary is the lamina of the leaf.

410. The style is an elongation of the midrib (208).

411. The stigma is the denuded, secreting, humid apex of the midrib.

412. Where the margins of a folded leaf, out of which the carpel is formed, meet and unite, a developement of cellular tissue sometimes takes place, forming what is called the *marginal placenta*.

413. Every such placenta is therefore composed of two parts, one of which belongs to one margin of the carpel, and one to the other.

414. But although the placenta of many plants appears to derive its origin from the margin of the carpels, it is certain that in many other instances the placenta is a mere developement of the centre of the flower-bud, and in reality the end of the medullary system. Such a placenta is called *central*.

It is not impossible that even marginal placentæ may be so in appearance only, and be in reality central.

415. This law will explain the structure of some anomalous pistils, in which the carpels are united into a confused mass; *Ex.* the Pomegranate[271].

416. As the carpels are modified leaves, they necessarily obey the laws of arrangement of leaves, and are therefore developed round a common axis.

417. And as they are leaves folded inwards, their margins are necessarily turned towards the axis. A placenta, therefore, formed by the union of those margins, will be invariably next the axis.

418. So that if a whorl of several carpels with a marginal placentation unite and constitute a pistil, the placentæ of that pistil will be all in the axis.

419. The normal position of the carpels is alternate with the innermost row of stamens, to which they are also equal in number; but this symmetry of arrangement is constantly destroyed by the abortion or non-developement of part of the carpels.

420. The carpels often occupy several whorls, in which case they are usually distinct from each other; *Ex.* Ranunculus, Fragaria, Rubus[272].

421. Sometimes, notwithstanding their occupying more than one whorl, they all unite in a single pistil; *Ex.* Nicotiana multivalvis, Monstrous Citrons. In these cases the placentæ of the innermost whorl of carpels occupy the axis, while those of the exterior carpels are united with the backs of the inner ones, as must necessarily happen in consequence of the invariable direction of the placentæ towards the axis.

422. When the carpels are arranged round a convex receptacle (389), the exterior ones will be lowest; *Ex.* Rubus[272].

423. But if they occupy the surface of a tube, or are placed upon a concave receptacle, the exterior ones will be uppermost; *Ex.* Rosa[177].

424. Whenever two carpels are developed, they are invariably opposite each other, and never side by side. This happens in consequence of the law of alternate opposition of leaves (191).

425. When carpels unite, those parts of their sides which are contiguous grow together, and form partitions between the cavities of the carpels.

E 2

426. These partitions are called *dissepiments*.

427. Each dissepiment is therefore formed of two layers. But these often grow together so intimately as to form but one layer.

428. Such being the origin of the dissepiments, it follows that,

429. All dissepiments are vertical, and never horizontal.

430. They are uniformly equal in number to the carpels out of which the pistil is formed.

431. A single carpel can have no dissepiment whatever.

432. It will also be apparent, that as the stigma must bear the same relation to the dissepiments as the point of the leaf to the sides of the lamina, the stigma will always be alternate with (between) the dissepiments.

433. When the dissepiments of a many-celled pistil are contracted so as not to separate the cavity into a number of distinct cells, but merely project into a cavity, the placentæ, which occupy the edges of these dissepiments, become what is called *parietal*; *Ex.* Poppy[q r]. Occasionally the placentæ are diffused over the whole face of the dissepiments, as in Butomus.

434. A one-celled ovary may also be formed out of several carpels, in consequence of the obliteration of dissepiments; *Ex.* Nut.

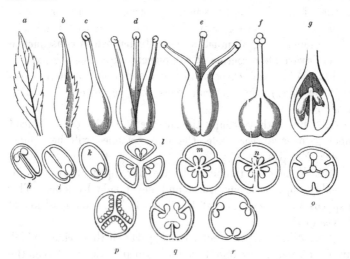

Some of the foregoing diagrams explain these laws : *a* is a leaf; *b*, a leaf rolled up preparatory to its conversion into a carpel ; *c* and *k*, a carpel ; *d* and *l*, three carpels approximated, but not united ; *e* and *m*, the same united at the ovaries, but disunited at the styles; *f* and *n*, these completely united into one ovary, one style, and one stigma.

435. All dissepiments whose position is at variance with the foregoing laws are spurious.

436. *Spurious dissepiments* derive their origin from various causes, and may have either a vertical or horizontal position.

437. When they are horizontal they are called *phragmata*, and are formed by the distension of the lining of the ovary ; *Ex.* Cathartocarpus, Fistula.

438. If vertical, they either are projections from the back of the carpel, as in Amelanchier and Thespesia[i] ; or they are caused by modifications of the placentæ, as in Martynia, Didymocarpus, and Cruciferæ ; or they are produced by the turning inwards of the margins of the carpels[k].

The singular fruit of Diplophractum[244], consisting of five cavities in the axis, surrounded by five two-celled cavities at the circumference, must be composed of carpels constructed as just described, and arranged in several series (420). This is explained by the following cut, where[244] is a section of the fruit of Diplophractum ; [242] shows an ideal arrangement of fifteen carpels in three rows, five being external and perfect, with the margins of the carpels turned inwards (406) ; five being altogether imperfect, and the five in the centre being less imperfect. [243] shows the transverse section of this ideal figure. In the ripe fruit we must suppose the intermediate carpels to be obliterated, and the spurious dissepiments of the external carpels to be pressed up against their back, so as to bisect the cavity of each carpel.

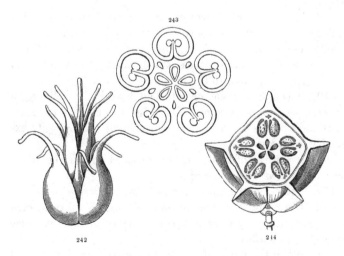

243

242   244

439. Sometimes the central placenta extends beyond the base of the carpels, rising up between them, and either forming an adhesion with the styles, as in Geranium, or a central distinct axis, as in Euphorbia.

440. This elongation of the placenta is more apparent in the fruit than in the pistil. It is analogous to the cellular apex of the spadix (304) of Arum.

441. The styles of different carpels frequently grow together into a solid cylinder*; *Ex.* Lilium. There are various degrees of union between the styles.

442. The style is incorrectly said to be divided in different ways, in consequence of this adhesion.

443. If the *ovary* adheres to the sides of the calyx it is called *inferior*, and the *calyx* is said to be *superior ; Ex.* Apple.

444. If it contracts no adhesion with the sides of the calyx it is called *superior*, and the *calyx inferior*.

XIV. OVULE.

445. The Ovule is a body borne by the placenta (412), and destined to become a seed (531).

446. It is to the carpel (406) what the marginal buds are to leaves (185), and to the central placenta what buds are to branches.

447. It does not, however, appear to bear any other analogy to a bud than what is indicated by its position.

448. The ovule is usually enclosed within an ovary (396); but in Coniferæ and Cycadaceæ it is destitute of any covering, and is exposed, naked, to the influence of the pollen.

449. It is either sessile, or attached by a little stalk called the *funiculus*, or *podosperm*. The point of union of the funiculus and ovule is the *base* of the latter, and the opposite extremity is its *apex*.

450. It consists of two sacs, one enclosed within the other, and of a *nucleus* within the sacs.

451. These sacs are called the *primine* and *secundine*.

452. The primine, secundine, and nucleus, are all connected with each other by a perfect continuity of tissue, at some point of their surface.

453. When the parts of the ovule undergo no alteration of

position during their growth, the two sacs and the nucleus are all connected at the base (449) of the ovule, which is *orthotropous* or *atropous*.

454. And then the base of the nucleus and that of the ovule are in immediate connection with each other.

455. But the relative position of the sacs and the base of the ovule are often entirely altered during the growth of the latter, so that it frequently happens that the point of union of the sacs and the nucleus is at the apex (449) of the ovule.

456. And then the base of the nucleus is at the apex of the ovule.

457. In such cases, a vascular connection is maintained between the base of the ovule and the base of the nucleus, by means of a bundle of vessels called a *raphe*.

458. The normal position of this raphe is on the side of the ovule, next the placenta.

459. The expansion of the raphe, where it communicates with the base of the nucleus, gives rise to the part of the seed called the *chalaza* (548).

460. When the ovule is curved downwards so as to approach the placenta, it is *campylotropous ;* when curved downwards and grown to the lower half, *anatropous ;* when attached by its middle so that the foramen is at one end and the base at the other, it is *amphitropous*.

461. The mouths of the primine and secundine usually contract into a small aperture called the *foramen* of the ovule, or the *exostome*.

462. The apex of the nucleus is always applied to this foramen.

463. In consequence of the relation the base of the nucleus bears to the base of the ovule, the foramen will be at the apex of the ovule when the two bases correspond, and at the base of the ovule when the two bases are diametrically opposite.

464. The foramen indicates the future position of the radicle of the embryo (555) ; the radicle being always next the foramen. This is a fact of great importance in practical Botany.

465. Within the nucleus is a cavity or bag, called the *sac*

of the amnios, containing a fluid named the liquor amnios, among which the embryo is developed.

XV. IMPREGNATION.

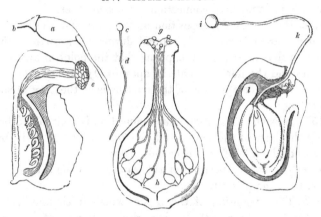

466. Impregnation is effected by contact between the pollen (378) and the stigma (397).

467. The pollen[c i] emits a tube[d k] of extreme delicacy, which pierces the stigma[e] and style[g], and, passing downwards into the ovary[f h], enters the foramen (461) of the ovule[a l].

468. Having reached the foramen, it comes into contact with the nucleus (450).

469. This accomplished, the act of impregnation is over; a new body gradually appears in the sac of the amnios (465), and eventually becomes an embryo.

470. Great numbers of modifications of this phenomenon have been observed, but they all resolve themselves into these facts.

471. In plants, the ovules of which have no pericarpial covering, such as Cycadaceæ and Coniferæ, (gymnosperms,) the pollen falls in the foramen, and there acts as if it had struck the stigma.

472. If only one pollen tube enters an ovule, there is but one embryo in the seed. But if several pollen tubes pass into the same ovule, there may be several embryos in the same seed; *Ex.* Onion, Miseltoe.

XVI. FRUIT.

473. The FRUIT, in the strictest sense of the word, is the pistil arrived at maturity. But the term is also applied to the pistil and floral envelopes taken together, whenever they are all united in one uniform mass.

474. Hence, whatever is the structure of the pistil, the same should be the structure of the fruit.

475. But in the course of the advance of the pistil towards maturity, many alterations take place, in consequence of abortion, non-developement, obliteration, and union of parts.

476. Whenever the fruit contains anything at variance with the laws that govern the structure of the pistil, the latter should be examined for the purpose of elucidation.

477. Sometimes a pistil with several cells produces a fruit with but one; *Ex.* the Hazel-nut and Cocoa-nut. This arises from the obliteration of part of the cells.

478. Or a pistil, consisting of one or two cells, changes to a fruit having several: the cause of this is a division and doubling of the placentary divisions; *Ex.* Martynia: or the expansion of portions of the interior; *Ex.* Cathartocarpus, Fistula.

479. As the fruit is the maturation of the pistil, it ought to indicate upon its surface some traces of a style; and this is true in all cases, except Cycadaceæ and Coniferæ, which have no ovary.

480. Hence the grains of corn, and many other bodies that resemble seeds, having traces of the remains of a style, cannot be seeds, but are minute fruits.

481. That part which was the ovary in the pistil, becomes the pericarp in the fruit.

482. The PERICARP consists of three parts; the outer coating called the *epicarp*, the inner lining called the *endocarp*, or *putamen*, and the intermediate substance named the *sarcocarp*.

483. Sometimes these three parts are all readily distinguished; *Ex.* the Peach: frequently they form one uniform substance; *Ex.* a Nut.

484. The *base* of the fruit is the part where it is joined to the peduncle. The *apex* is where the remains of the style are found.

485. The axis of the fruit is often called the *columella*; the space where two carpels unite is named the *commissure*.

486. All fruits which are mere modifications of a single carpellary leaf (406) have always a suture corresponding with the junction of the margins, or with the placentæ, and often another corresponding with the midrib of the carpellary leaf: the former is called the *ventral*, the latter the *dorsal suture*.

487. If the pericarp neither splits nor opens when ripe, it is said to be *indehiscent*; if it does split or open, it is said to *dehisce*, or to be *dehiscent*; and the pieces into which it splits are called the *valves*.

488. The dehiscence of the pericarp takes place in different ways.

489. If it takes place longitudinally, or vertically, so that the line of dehiscence corresponds with the junction of the carpels, the dissepiments are divided, the cells remain closed at the back, and the *dehiscence* is called *septicidal*; *Ex*. Rhododendron[264].

490. Formerly, botanists said that in this kind of dehiscence the *valves were alternate with the dissepiment*; or, that the *valves had their margins turned inwards*.

491. If it takes place vertically, so that the line of dehiscence corresponds with the dorsal suture (486), the dissepiments remain united, the cells are opened at their back, and the dehiscence is called *loculicidal*; *Ex*. Lilac, Lily.

492. Formerly, it was said that in this kind of dehiscence *the dissepiments were opposite the valves*.

493. When a separation in the pericarp takes place across the cells horizontally, the dehiscence is *transverse*; *Ex*. Anagallis.

494. If the dehiscence is effected by partial openings of the pericarp, it is said to take place by pores; *Ex*. Poppy.

495. Sometimes the cells remain closed, separating from the axis formed by the extension of the peduncle (284); *Ex*. Umbelliferæ, Euphorbia[255].

496. Or the cells open and separate from the axis, which is formed by a cohesion of the placentæ which separate from the dissepiments; *Ex*. Rhododendron[264].

497. Sometimes the dissepiments cohere at the axis, and separate from the valves (487) or back of the carpels; *Ex*. Convolvulus.

498. All fruits are either *simple* or *multiple*.

499. Simple fruits proceed from a single flower; *Ex.* Pæony, Apple, Nut, Strawberry.

500. Multiple fruits are formed out of several flowers [D F]; *Ex.* Fir, Pine-apple, Fig. They are masses of inflorescence in a state of adhesion, and are also called *anthocarpous*.

501. Simple fruits are either the maturation of a single carpel (406), or of a pistil formed by the union of several carpels (408).

502. Of fruits formed of a single carpel, the most important are the Follicle (503), Legume (504), Drupe (507), Achenium (508), Caryopsis (511), and Utricle (512).

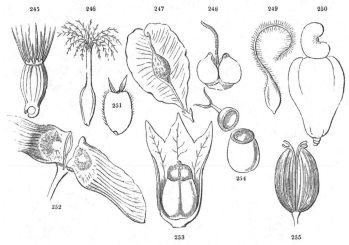

503. The *Follicle* is a carpel dehiscing by the ventral suture, and having no dorsal suture[260].

504. The *Legume* is a carpel having both a ventral and dorsal suture, and dehiscing by both, either, or neither[263 E 258].

505. The two sutures of a legume sometimes form what is called a *replum*; *Ex.* Carmichælia.

506. When articulations take place across the legume, and it falls into several pieces, it is said to be *lomentaceous*[257 262].

507. The *Drupe* differs from the follicle in being indehiscent, and in its pericarp having a distinct separation of epicarp (482), sarcocarp, and endocarp[259].

508. The *Achenium* is an indehiscent, bony, one-seeded

pericarp, which does not contract any degree of adhesion with the integument of the seed[251] [249].

509. It is a drupe, the pericarp of which does not separate into three layers.

> The Achenium is pappose when it bears the remains of a calyx at its apex ; Ex. Compositæ : and is truncate[245], or rostrate[246], while the pappus is setaceous[245], double[245], plumose[246], or paleaceous[251]. If the style remains and becomes feathery, forming a kind of tail, the achenium is caudate[249].

510. Occasionally the achenium is elevated on a large fleshy receptacle, as in Anacardium[250].

511. The *Caryopsis* is an indehiscent, membranous, one-seeded pericarp, which adheres firmly to the integument of the seed ; *Ex.* Corn.

512. The *Utricle* is a caryopsis, the pericarp of which has no adhesion with the integuments of the seed ; *Ex.* Eleusine, Chenopodium.

513. Of fruit formed of several carpels, the principal are the Capsule (514), Pyxis (520), Samara (517), Cremocarp (518), Nuculanium (519), Siliqua (515), Nut or Gland (517), Berry (522), Orange (523), Pome (524), Pepo (525), and Balausta (526).

514. The *Capsule* is a many-celled, dry, dehiscent pericarp[253] [256] [264] [269].

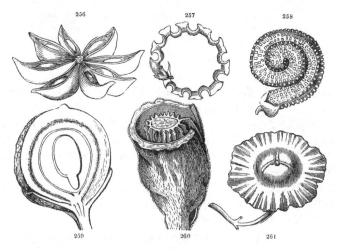

> It is stellate[256], toothed at the apex[265], or spiral[267] ; if its cells remain close after separation[268], they are named cocci.

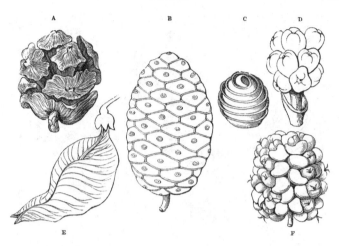

515. The *Siliqua* consists of two carpels fastened together, the placentæ of which are parietal, and separate from the valves, remaining in the form of a replum (505), and connected by a membranous expansion[266].

516. When the siliqua is very short, or broader than it is long, it is called a *Silicula*.

517. The *Nut* or *Gland* is a dry, bony, indehiscent, one-celled fruit, proceeding from a pistil of three cells, and enclosed in an involucre called a *Cupule* ; *Ex.* the Hazel, Acorn. It is a sort of compound achenium.

> In some Palms, *Ex.* Sagus, it is covered by scales turned downwards[273]. It is often bordered by expansions or wings which surround it longitudinally, as in the Elm[247] ; or transversely, as in Paliurus[261] ; or proceed from the apex or back only, as in Sycamore[252], in which case it receives the name of *Samara*.

518. The *Cremocarp* is a pair of Achenia, then called *mericarps*, placed face to face, and separating from a central axis ; *Ex.* Umbelliferæ[255]. Their planes of union constitute the *commissure*.

519. The *Nuculanium* is a capsule, which, being fleshy, does not dehisce ; *Ex.* Grape, Arbutus[270].

520. The *Pyxis* is a capsule whose dehiscence takes place transversely[253] [254] ; *Ex.* Hyoscyamus, Anagallis.

521. The *Etærio* is a collection of distinct, indehiscent carpels, fleshy or dry, within a calyx ; *Ex.* Rubus[272].

522. The *Berry* is a succulent fruit, the seeds of which lose their adhesion when ripe, and lie loose in pulp ; *Ex.* a Gooseberry.

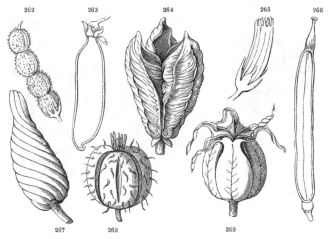

523. The *Orange* is a berry having a pericarp separable into an epicarp, an endocarp, and a sarcocarp, and the cells filled with pulpy bags, which are cellular extensions of the sides of the cavity.

524. The *Pome* is a union of two or more inferior carpels, the pericarp being fleshy, and formed of the floral envelope and ovary firmly united[274].

525. The *Pepo* is composed of about three carpels, forming a three-celled, fleshy, indehiscent fruit, with parietal placentæ; *Ex.* Cucumber.

526. The *Balausta* is a many-celled fruit, with the seeds

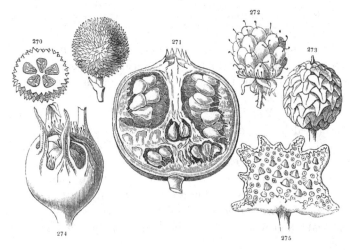

arranged in an irregular manner on the backs of the cells, and is formed by more whorls of carpels than one, enclosed within a tough rind; *Ex.* Pomegranate[271].

527. The most remarkable modifications of multiple or anthocarpous fruits are, the Cone (528), Pine-apple (529), and Fig (530).

528. The *Cone* [A B] is an indurated amentum (305); *Ex.* Pinus. When it is much reduced in size, and its scales firmly cohere, it is called a *Galbulus; Ex.* Thuja.

529. The *Pine-apple* is a spike of inferior flowers, which all grow together into a fleshy mass.

530. The *Fig* is the fleshy, hollow, dilated apex of a peduncle, within which a number of flowers are arranged, each of which contains an achenium; *Ex.* Ficus, Dorstenia[275].

Of the terms above explained only a few are in common use, and it seems to be found by systematic botanists more convenient to describe a given fruit by exact words than to use any particular term. The names most employed are the Achenium, Nut, Caryopsis, Drupe, Capsule, Siliqua, Legume, and Cone.

XVII. SEED.

531. The SEED is the ovule (406) arrived at maturity.

532. It consists of integuments (540), albumen (551), and embryo (555); and is the result of the reciprocal action of the sexual apparatus.

533. In general, seeds are, like ovules, enclosed within a covering arising from a carpellary leaf (406); but all Gymnosperms are an exception to this. Moreover, some ovules rupture the ovary soon after they begin to advance towards the state of seed, and thus become naked seeds; *Ex.* Leontice. Others are imperfectly protected by the ovary, the carpels not being perfectly closed up; *Ex.* Reseda.

534. The seed proceeds from the placenta (412), to which it is attached by the funiculus[280], which is sometimes very long, but is more frequently not distinguishable from the placenta.

535. Sometimes the funiculus, or the placenta, expands about the seed into a fleshy body; *Ex.* the *Mace* of a nutmeg, Euonymus. This expansion is named *aril*[276 281 283].

536. It is never developed until after the vivification of the ovule, and must not be confounded with tumours or dilatations of the integument of the seed.

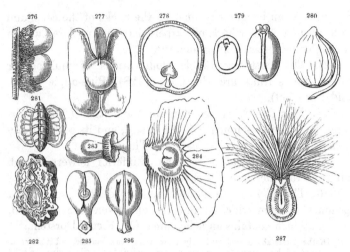

537. Sometimes there are tumours of the testa near the hilum or at the opposite end; such are called *Strophiolæ* or *Carunculæ*[279].

538. The precise nature of these is unknown; sometimes they are dilatations of the chalaza; *Ex.* Crocus: or they are caused by a fungous state of the lips of the foramen; *Ex.* Ricinus: or they arise from unknown causes.

539. The scar, which indicates the union of the seed with the placenta, is called the *hilum* or *umbilicus*[285].

540. The integuments are called collectively *testa*, and consist of membranes resulting from the sacs of the ovule (451).

541. Sometimes the testa is covered by hair-like expansions of its whole surface; as in the Cotton: or these hairs occupy one or both ends, when they constitute what is called the *coma*[287]. This must not be confounded with pappus (328), which is calyx.

542. The integuments are often expanded into wings, which are either single[284] or several[277], and appear intended to render seeds buoyant. Very often they are corky or spongy[282], and not unfrequently consist of spiral cells (19).

543. In the seed these membranes are called by various names, of which the most frequently used are *spermoderm* or *testa* for the primine; *mesosperm*, for the secundine; and *endopleura* for the coat of the nucleus (450).

544. The mouth of the foramen (461) is often distinctly visible, and is named the *micropyle; Ex.* Pea.

545. The *raphe*[279] [285] occupies one side of the seed in all cases in which it pre-existed in the primine; but it frequently becomes much ramified.

546. The raphe is in no way connected with impregnation; its functions being apparently confined to maintaining a vascular connection between the placenta and the base of the nucleus, for the purpose of nourishing the latter.

547. Spiral vessels are found in the raphe and its ramifications.

548. Where vessels of the raphe expand into the mesosperm (543), the *chalaza* (459) appears as a discoloured thickening of the integuments[285].

549. The micropyle always indicates the point in the circumference of a seed towards which the radicle (561) points.

550. And the chalaza is as constant an indication, when it is present, of the situation of the cotyledons (559); it being always at that part of the circumference organically opposed to the radicle.

551. Between the integuments and the embryo of some plants lies a substance called the *albumen* or *perisperm*[278] [293] [299].

552. It consists of a peculiar matter deposited during the growth of the ovule among the celullar tissue of the nucleus (450).

553. When the cellular tissue of the nucleus combines with the deposited matter so completely as to form together but one substance, the albumen is called solid; *Ex.* Wheat, Euphorbia. When a portion of the tissue remains unconverted, the albumen is *ruminated; Ex.* Anona, Nutmeg.

554. Albumen is usually wholesome, and may be frequently eaten with impunity in the most dangerous tribes; *Ex.* Omphalococca, a genus of Euphorbiaceæ.

555. The organised body that lies within the seed, and for the purpose of protecting and nourishing which the seed was created, is the *Embryo*[278].

556. The embryo was originally included within the sac of the amnios (465).

557. The latter is usually absorbed or obliterated during

F

the advance of the embryo to maturity; but it sometimes remains surrounding the ripe embryo, in the form of *Vitellus*; *Ex.* Saururus, Piper[297].

558. The embryo consists of the cotyledons (559), the radicle (561), the plumule (560), and the collar (562).

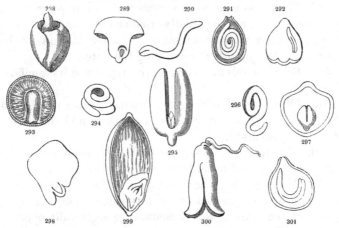

559. The *cotyledons* represent undeveloped leaves[295a].

560. The *plumule*, or *gemmule*, is the nascent ascending axis (64)[295b].

561. The *radicle* is the rudiment of the descending axis (71)[295c].

562. The *collar* is the line of separation between the radicle and the cotyledons.

563. The space that intervenes between the collar and the base of the cotyledons is called the *cauliculus*. (Tigelle, *Fr.*)

564. In some seeds the embryo is furnished with a *suspensor* from the point of the radicle[300].

565. The embryo is usually solitary in the seed, but occasionally there are two or several (472).

566. When several embryos are produced within a single seed, it sometimes happens that two of these embryos grow together, in which case a production analogous to animal dicephalous monsters is formed.

In form, position, and direction, the embryo varies in different species. In general it is straight; in some it is spiral[291]; in others heliacal[294]; in others vermicular[290]; in others arcuate[301]. It usually occupies the axis of the albumen or seed[278] [293]: but it is also excentrical[292]; and unilateral[299]. In direction, it is either erect with respect to the seed, or inverted or transverse.

567. The number of cotyledons varies from one to several. The most common number is either one or two. In the latter case, they are always directly opposite each other.

The cotyledons are semiterete[295]a; foliaceous[278]; flat, convolute[288]; parallel with each other, or divergent[300]. When there is but one cotyledon, it often assumes peculiar forms: it is, for instance, fungous[289]; spheroidal[298]a; lenticular[299]a.

568. The direction of the embryo, with respect to the seed, will depend upon the relation that the integuments, the raphe, chalaza, hilum, and micropyle, bear to each other.

569. If the nucleus be inverted, the embryo will be erect, or *orthotropous*; *Ex.* Apple.

570. If the nucleus be erect, the embryo will be inverted, or *antitropous*; *Ex.* Nettle.

571. If the micropyle is at neither end of the seed, the embryo will be neither erect nor inverted, but will be in a more or less oblique direction with respect to the seed; *Ex.* Primrose; and is said to be *heterotropous*.

572. Plants that have but one cotyledon, or, if two, with the cotyledons alternate with each other, are called MONO-COTYLEDONOUS[293 289 298 299].

573. Plants that have two opposite each other, or a greater number placed in a whorl, are called DICOTYLEDONOUS[288 290 292 297 300 301].

574. Endogenous plants are monocotyledonous.

575. Exogenous plants are dicotyledonous.

576. Plants that have no cotyledons are said to be ACOTYLE-DONOUS[294].

577. But this term is usually applied only to cellular plants which, having no sexual apparatus, can have no seeds (587).

578. Acrogenous plants are acotyledonous.

579. Those seeds of flowering plants, which appear to have no cotyledons, owe their appearance to the cotyledons being consolidated; *Ex.* Lecythis, Olynthia: or abortive; *Ex.* Cuscuta.

580. The plumule is very often latent, until it is called into action by the germination of the seed. Sometimes it is undistinguishable from the cotyledons; sometimes it is highly developed, and lies in a furrow of the cotyledon; *Ex.* Maize

[299]. In the monocotyledonous embryo it frequently happens that the plumule is rolled up in the cotyledon, the margins of which grow together, so that the whole embryo forms one uniform mass[293]; but as soon as germination commences the margins separate.

581. The radicle elongates downwards, either directly from the base of the embryo, or after previously rupturing the integument of the base. Plants with the first character are called Exorhizæ[295]; with the second, Endorhizæ[298] [299].

582. The endorhizal embryo is very common in monocotyledons; the exorhizal, in dicotyledons.

583. When the seed is called into action, germination takes place. The juices of the plant, which before were insipid, immediately afterwards abound with sugar; *Ex.* Barley; and growth commences.

584. This growth is in the first instance caused by the absorption and decomposition of water, whose oxygen combines with the superfluous carbon of the seed, and is expelled in the form of carbonic acid gas.

585. As this phenomenon does not take place in full-grown plants, except in the dark (258), so neither can it occur in seeds, except under the same condition. Hence an embryo, exposed to constant light, would not germinate at all; and hence the care taken by nature to provide a covering to all embryos in the form of the integuments of the seed or of a pericarp.

586. As soon as the necessary proportion of carbon is removed from a seed by the expulsion of carbonic acid, the young plant begins to absorb food, and to grow by the processes of assimilation and respiration already described (254).

ACROGENS, OR FLOWERLESS PLANTS.

587. Many plants not being increased by seeds, the result of the mutual action of sexual apparatus (531), are flowerless, and destitute of organs of fructification.

588. Such are propagated by what are called organs of reproduction, which have no other analogy with the organs of fructification than that both perpetuate the species.

589. The reproductive organs of flowerless plants vary according to the tribes of that division of the vegetable kingdom ; and have so little relation to each other, that each principal tribe may be said to have its own peculiar method of propagation.

590. They all agree in their reproductive parts or *spores*, which are analogous to seeds, not germinating from any fixed point, but producing root or stem indifferently from any point of their surface. This germination is therefore *vague*.

591. The principal tribes are *Ferns* (592), *Mosses* (598), *Lichens* (605), *Algaceæ* (607), and *Fungaceæ* (610).

592. FERNS are increased by little bodies, called *spores*, enclosed within cases named *thecæ* or *sporangia*[302] [303], which often grow in clusters or *sori*[304], from the veins of the under sides of the leaves, or from beneath the epidermis. The latter, when it encloses the *thecæ*, is termed the *indusium*[308].

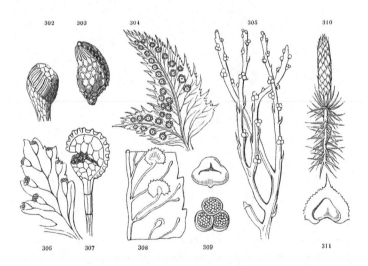

593. The indusium separates from the leaf in various ways, in consequence of the growth of the thecæ beneath it.

594. The thecæ have frequently a stalk which passes up one side, and finally, curving with their curvature, disappears on the opposite side[307].

595. The part where the stalk of the theca is united with its side, is called the *annulus*.

596. These thecæ may be considered minute leaves, having the same gyrate mode of developement as the ordinary leaves of the tribe; their stalk the petiole, the annulus the midrib, and the thecæ itself the lamina, the edges of which are united.

597. They would, therefore, be analogous to carpels, if it appeared that they were influenced by the action of any vivifying matter.

598. Mosses are increased by spores (590), contained within an *urn*, or *theca*, or *sporangium*[314] [316], placed at the apex of a stalk or *seta*, bearing on its summit a kind of loose hood, called a *calyptra*[314], and closed by a lid or *operculum*.

599. The inside of the theca has a central axis or *columella*, and the orifice beneath the operculum is closed by teeth-like processes, or a membrane called the *peristome*[313] [318].

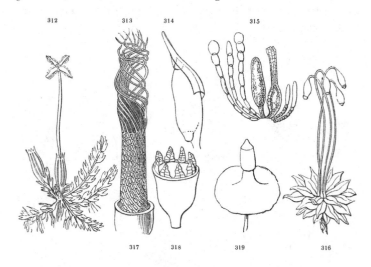

600. At the base of the theca is sometimes found a tumour or *struma*[314], or an equal expansion named *apophysis*[319].

601. The number of the teeth of the peristome is always some multiple of four.

602. The calyptra originally grew from the base of the stalk; but when the stalk lengthened, the calyptra was torn away from its base and carried up, surrounding the theca.

603. The calyptra may be understood to be a convolute leaf; the operculum, another; the peristome, one or more

whorls of minute flat leaves; and the theca itself to be the excavated distended apex of the stalk, the cellular substance of which separates in the form of sporules.

604. There are also in mosses certain organs, called anthers by some, which do not appear analogous to the male apparatus of flowering plants, and the nature of which has not been demonstrated. They are jointed filaments, *staminidia* or *antheridia*, containing vibrios lodged in mucous cells, and surround the rudiment of the future theca.

At figure 315 the flask-like figure is a young sporangium, or in this state pistillidium; and the club-shaped body on its left, a staminidium. The articulated threads may be abortive staminidia.

605. LICHENS are cellular expansions, usually horizontal, but occasionally perpendicular, consisting of a *thallus*[331], or combination of stem and leaves, upon which *shields, apothecia,* or reproductive organs, appear[331].

606. The shields consist of a margin, enclosing a kernel, *nucleus*, in which tubes containing sporules, and called *asci*, are imbedded.

They vary a little in nature, whence they have received the following other names: *scutellum*[337]; *orbilla*, which is the same thing; *pelta*[329]; *tuberculum*[335]; *trica* or *gyroma*; if covered with sinuous concentric furrows, *lirella*[333]; *patellula*[334]. Be-

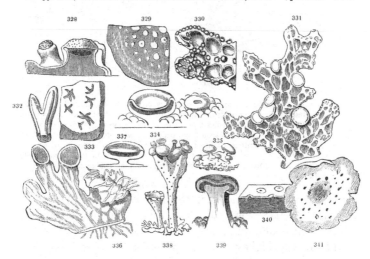

sides the foregoing, some other peculiar terms are used by writers on Lichens. *Asci* are tubes of the nucleus, containing sporules; the latter are sometimes named *gongyli: perithecium* is the part in which asci are immersed; *hypothecium* is a substance overlying the *perithecium*. *Podetia* are stalk-like elongations of the thallus[339]; *scypha* or *oplarium* is a cup-like expansion of a podetium, having shields on the margin[338]. *Soredia*, or powdery masses[330] ; they are also called *globuli* and *glomeruli*. *Lacunæ* are pits of the thallus[331]. *Excipulus* is that part of the thallus which forms the rim and base of shields. *Thallodes* signifies formed of the thallus.

607. ALGACEÆ are submersed plants, equally destitute of any kind of tissue, except the cellular, and propagated by spores (590) lodged in various parts of the system.

608. The sporules either lie freely in the whole substance of such plants, or are collected in particular cells[322], or occupy jointed filaments[324], or are placed in spheres[321], occupying the circumferences of expansions of the thallus (605).

609. There are also other modes of multiplication.

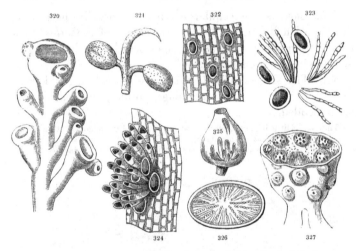

Among the special terms employed by writers on this order, the following may be enumerated as the principal. Among their reproductive organs are *gongyli*, or hard round deciduous bodies ; *granula*, or large spores ; *sporidia*, or bodies resembling spores, but not such ; *sporangia* or *coniocysta*, or spore-cases[321][325]. *Hypha* is a filamentous thallus ; *phycomater* is the gelatine in which the spores of some begin to vegetate ; *peridiolum* is a membrane immediately covering the spores ; *vesiculæ* are air-bladders that enable some species to float.

610. FUNGACEÆ, which are the lowest form of vegetation, are also cellular, some of their cells however containing spiral threads, and are propagated by spores.

611. In the highest forms, two kinds of organs are detected : one, *cystidia*[345], are conical naked elevations ; the other, *basidia*[345], are also conical elevations, but they bear spores in definite number on their apex.

612. The highest forms of the fungaceous order consist of a *stipes*[346], an *annulus* or *collar*[346], a *pileus*[346] or cap, and an *hymenium*.

613. Lower forms are reduced to a mere *peridium* or integument, containing the reproductive system[342].

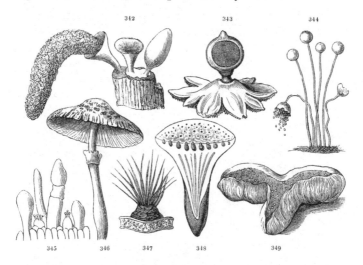

614. Some have the sporules enclosed in asci (606).

615. The lowest consist of nothing but cells, placed end to end, and enclosing spores in the terminal cells[344].

Of the special terms employed by Mycologists (writers on Fungaceous plants), the principal are the following. The *volva* is the wrapper which covers over many of them, as Agarics, in their youngest state. *Thallus* is the spawn usually generated under ground, or amongst decaying matter. *Velum* is a membrane that connects the pileus and collar. *Cortina* is that part of a velum which adheres to the margin of the pileus. *Flocci* are wool-like threads found mixed with sporules ; and *stroma* is the body on which flocci grow. *Orbiculi* are little disks contained within the peridia of certain genera. *Sporangium* is the external coating of such genera as Lycoperdon[343]. *Perithecium* is the bag of fructification in Sphæria[348]. *Ostiolum* is the mouth of the bag. *Capillitium* is a kind of purse or net containing spores[342]. *Mycelia* are nascent fungi, or fragments of their spawn.

II.—SYSTEMATICAL BOTANY.

616. SYSTEMATICAL BOTANY is the science of arranging plants in such a manner that their names may be ascertained, their affinities determined, their true place in a natural system fixed, their sensible properties judged of, and their whole history elucidated with certainty and accuracy.

617. Anything short of this is not a system, but an artificial scheme.

618. The latter is intended to enable a person to ascertain the name of a plant, and goes no further.

619. But as the name of a plant conveys no information by itself, the power thus acquired by artificial schemes is of but little real value, and cannot be considered as anything beyond a very imperfect and elementary mode of investigation.

620. What knowledge is gained by the use of an artificial scheme is a mere collection of isolated facts, without mutual dependence, or any distinct bearing upon general views.

621. In a natural arrangement, on the other hand, the name of a plant is the least object that is gained. Any investigation upon its principles, when completed, is, of necessity, attended with the discovery of the relationship a given plant bears to others; and as plants which are most closely akin in structure are also most similar in their sensible properties, it often enables us to judge of the use of an unknown plant whose place is determined in the system, by the ascertained uses of those species in whose vicinity it takes its place by virtue of its natural affinities.

622. The only artificial schemes in general use are, 1, that of Linnæus (623), called the SEXUAL System, in consequence of its characters being dependent upon variations in the stamens and pistil, or sexes, of plants; and 2, the Analytical method.

I. LINNÆAN SEXUAL SYSTEM.

623. This is now disused by men of science ; but, as many books still employed have been arranged upon its plan, it is necessary for a student to understand it.

624. Its divisions, called classes and orders, depend upon modifications of the stamens and pistils, and have Greek names expressive of their distinctive characters.

Class 1. Monandria. *Stam.* 1.
    2. Diandria. *Stam.* 2.
    3. Triandria. *Stam.* 3.
    4. Tetrandria. *Stam.* 4.
    5. Pentandria. *Stam.* 5.
    6. Hexandria. *Stam.* 6.
    7. Heptandria. *Stam.* 7.
    8. Octandria. *Stam.* 8.
    9. Enneandria. *Stam.* 9.
    10. Decandria. *Stam.* 10.
    11. Dodecandria. *Stam.* 12—19.
    12. Icosandria. *Stam.* 20 or more, perigynous (359).
    13. Polyandria. *Stam.* 20 or more, hypogynous (358).

Orders. Each of these classes is divided into orders characterized by the number of styles or sessile stigmas. Monogynia signifies 1 style; Digynia, 2; Trigynia, 3; Tetragynia, 4; Pentagynia, 5; Hexagynia, 6; Heptagynia, 7; Octogynia, 8; Enneagynia, 9; Decagynia, 10; Dodecagynia, &c. about 12; Polygynia, many.

Class 14. Didynamia: *Stamens* 4, two long and two short.
    Orders: 1. Gymnospermia, seeds apparently naked ; 2. Angiospermia, seeds evidently in a seed-vessel.

Class 15. Tetradynamia: *Stamens* 6, four long and two short.
    Orders: 1. Siliquosa, with a long pod; 2. Siliculosa, with a short pod or pouch.

Class 16. Monadelphia: Filaments united into a cup or column. *Orders:* 1. Pentandria; 2. Decandria, &c. as before.

Class 17. Diadelphia: Filaments united into two parcels or fraternities. *Orders:* 1. Hexandria, &c. as before.

Class 18. Polyadelphia: Filaments united into more parcels than two. *Orders:* 1. Dodecandria; 2. Icosandria, &c. as before.

Class 19. Syngenesia: *Anthers* united into a tube. *Orders:* 1. Monogamia, flowers solitary; 2. Polygamia, flowers in heads. *Sub-orders* of the latter: 1. Æqualis, florets all equal; 2. Superflua, florets of the disk complete, of the ray female; 3. Frustranea, florets of the disk perfect, of the ray neuter; 4. Necessaria, florets of the disk male, of the ray female; 5. Segregata, florets each with its own proper involucre.

Class 20. Gynandria: Stamens and styles consolidated. *Orders:* 1. Monandria, &c. as before.

Class 21. Monœcia: Stamens in one flower, pistils in another, on the same plant. *Orders:* 1. Monandria, &c. as before.

Class 22. Diœcia: Stamens in one flower, pistils in another, on different plants. *Orders:* 1. Monandria, &c. as before.

Class 23. Polygamia: Stamens and pistils separate in some flowers, united in others, either on the same plant or on two or three different ones. *Orders:* 1. Monœcia, &c. as before.

Class 24. Cryptogamia: no apparent flowers. *Orders:* Filices, Musci, Hepaticæ, Algæ, Fungi (592).

II. ANALYTICAL METHOD.

625. This is founded upon the common process of analysis that is unconsciously employed by the human mind. In all cases the mental operation by which one thing is distinguished from another, consists in a continual contrast of characters. For instance, in a mass of individuals we distinguish one set which is coloured, and another which is colourless; of those that are coloured we distinguish red, black, blue, and green; of the red, some are square, others are round; of the round, some are sculptured on their surface, others are even:—and so we proceed, analysing the subject by a constant series of contrasts, until we have arrived at a point beyond which no analysis can go.

626. The following pages contain such an analysis of the principal natural orders of plants. The method may be equally applied to genera and species, and is an instructive process if employed by way of exercise to the mind, and for the purpose of rendering distinctions definite:

1. Plants with visible flowers and seeds 2
Plants with visible flowers and spores . . . RHIZANTHS.
Plants without flowers . . 261
2. Leaves netted. Wood in concentric layers . . . 3
Leaves straight-veined. Wood confused . : . . 239
Leaves straight-veined. Wood in concentric layers . . 237
3. Flowers having both calyx and corolla 4
Flowers having a calyx only, or none 119
4. Petals distinct . . . 5
Petals united . . . 186
5. Stamens more than 20 . . 6
Stamens fewer than 20 . 36
6. Ovary inferior, or partially so . 7
Ovary superior . . . 15
7. Leaves furnished with stipules . 8
Leaves without stipules . . 10
8. Carpels more or less disunited POMEÆ.
Carpels consolidated . . 9

9. Placenta central LECYTHIDACEÆ.
Placenta parietal . HOMALIACEÆ.
10. Carpels distinct . ANONACEÆ.
Carpels consolidated . . 11
11. Placenta spread over the septa NYMPHÆACEÆ.
Placenta parietal · . 12
Placenta in the axis . . 13
12. Petals definite in number LOASACEÆ.
Petals indefinite . CACTACEÆ.
13. Leaves with transparent dots MYRTACEÆ.
Leaves dotless . . 14
14. Petals indefinite . MESEMBRYACEÆ.
Petals definite . PHILADELPHACEÆ.
15. Leaves with stipules . . 16
Leaves without stipules . . 24
16. Carpels disunited . . 17
Carpels consolidated . . 18
17. Stamens hypogynous MAGNOLIACEÆ.
Stamens perigynous . ROSACEÆ.
18. Placenta parietal . BIXACEÆ.
Placenta in the axis . . 19
19. Æstivation of calyx imbricated . 20
Æstivation valvate . . 22

20. Flowers unisexual   EUPHORBIACEÆ.
     Flowers hermaphrodite .   . 21
21. Ovary one-celled . PORTULACACEÆ.
     Ovary with more cells than one
                     CISTACEÆ.
22. Calyx enlarged and irregular
                  DIPTERACEÆ.
     Calyx not enlarged .   . 23
23. Stamens monadelphous   MALVACEÆ.
     Stamens distinct . TILIACEÆ.
24. Carpels disunited or solitary . 25
     Carpels consolidated .   . 30
25. Carpels plunged in a tabular disk
               NELUMBIACEÆ.
     Carpels clear of the disk . 26
26. Stamens perigynous . ROSACEÆ.
     Stamens hypogynous .   . 27
27. Carpel solitary . ANACARDIACEÆ.
     Carpels several .   .   . 28
28. Stamens polyadelphous HYPERICACEÆ.
     Stamens free .   .   . 29
29. Herbs .   . RANUNCULACEÆ.
     Trees or Shrubs . ANONACEÆ.
30. Placentas in the axis .   . 31
     Placentas parietal .   . 34
     Placentas dissepimental
               NYMPHÆACEÆ.
31. Stigma large, umbrella-shaped
               SARRACENIACEÆ.
     Stigma simple .   .   . 32
32. Sepals 2 .   PORTULACACEÆ.
     Sepals more than 2 .   . 33
     Sepals united into a tube LYTHRACEÆ.
33. Petals flat, seeds few, leaves
     leathery .   .   . CLUSIACEÆ.
     Petals crumpled, seeds numerous,
     leaves membranous . CISTACEÆ.
     Petals flat, stamens monadel-
     phous .   .   . HUMIRIACEÆ.
34. Placentas spread over the lining
     of the fruit . FLACOURTIACEÆ.
     Placentas in lines .   . 35
35. Ovary stalked . CAPPARIDACEÆ.
     Ovary sessile . PAPAVERACEÆ.
36. Ovary more or less inferior . 37
     Ovary quite superior .   . 55
37. Leaves with stipules .   . 38
     Leaves without stipules .   . 41
38. Flowers unisexual . BEGONIACEÆ.
     Flowers hermaphrodite .   . 39

39. Placentas parietal . HOMALIACEÆ.
     Placentas in the axis .   . 40
40. Calyx valvate, stamens opposite
     the petals .   . RHAMNACEÆ.
     Calyx imbricate, stamens alternate
     with petals . HAMAMELACEÆ.
41. Placentas parietal .   . 42
     Placentas axile .   .   . 43
42. Flowers unisexual   CUCURBITACEÆ.
     Flowers hermaphr. or polyg.
               GROSSULACEÆ.
43. Disk double .   . APIACEÆ.
     Disk simple .   .   . 44
44. Seeds few .   .   . 45
     Seeds numerous .   . 51
45. Carpels solitary .   .   . 46
     Carpels several .   .   . 48
46. Parasites on trees . LORANTHACEÆ.
     Root plants .   .   . 47
47. Leaves balsamic, acrid
               ANACARDIACEÆ.
     Leaves insipid . COMBRETACEÆ.
48. Calyx valvate .   .   . 49
     Calyx imbricated .   . 50
49. Stamens opp. petals . RHAMNACEÆ.
     Stamens altern. petals . CORNACEÆ.
50. Anthers curved downwards
               MEMECYLACEÆ.
     Anthers erect . BRUNIACEÆ.
51. Leaves dotted . MYRTACEÆ.
     Leaves not dotted .   . 52
52. Anthers curved downwards
               MELASTOMACEÆ.
     Anthers erect .   .   . 53
53. Flowers tetramerous . ONAGRACEÆ.
     Flowers not tetramerous . 54
54. Petals always distinct
               SAXIFRAGACEÆ.
     Petals at first united into a tube
               ESCALLONIACEÆ.
55. Leaves with stipules .   . 56
     Leaves without stipules . 81
56. Carpels disunited .   . 57
     Carpels consolidated .   . 59
57. Anther valves recurved
               BERBERACEÆ.
     Anther valves straight .   . 58
58. Fruit leguminous . FABACEÆ.
     Fruit drupaceous or capsular
               ROSACEÆ.

59. Placentas parietal   .   . 60
    Placentas in the axis   .   . 62
60. Flowers with a coronet
                          PASSIFLORACEÆ.
    Flowers without a coronet   . 61
61. Leaves circinate when young
                          DROSERACEÆ.
    Leaves straight when young
                          VIOLACEÆ.
62. Flowers unisexual   EUPHORBIACEÆ.
    Flowers hermaphrodite or polyg.   63
63. Sepals 2 .   . PORTULACACEÆ.
    Sepals more than 2   .   . 64
64. Fruit with a long beak
                          GERANIACEÆ.
    Fruit without a beak   .   . 65
65. Styles distinct to the base   . 66
    Styles more or less united   . 72
66. Petals minute .   .   . 68
    Petals conspicuous   .   . 69
68. Stigmas capitate   . ELATINACEÆ.
    Stigmas simple   . ILLECEBRACEÆ.
69. Calyx valvate   . ELÆOCARPACEÆ.
    Calyx imbricated   .   . 70
70. Stamens hypogynous
                          MALPIGHIACEÆ.
    Stamens perigynous   .   . 71
71. Leaves opposite   . CUNONIACEÆ.
    Leaves alternate . SAXIFRAGACEÆ.
72. Calyx imbricated   .   . 73
    Calyx valvate or open .   . 77
73. Stamens monadelphous OXALIDACEÆ.
    Stamens distinct   .   . 74
74. Calyx surrounded with double
    glands   .   . MALPIGHIACEÆ.
    Calyx simple .   .   . 75
75. Leaves simple   ZYGOPHYLLACEÆ.
    Leaves compound   .   . 76
76. Flowers unsymmetrical SAPINDACEÆ.
    Flowers symmetrical
                          STAPHYLEACEÆ.
77. Stamens opposite the petals   . 78
    Stamens alternate with petals   . 79
78. Stamens perigynous . RHAMNACEÆ.
    Stamens hypogynous   . VITACEÆ.
79. Anthers opening by pores
                          ELÆOCARPACEÆ.
    Anthers opening by slits   . 80
80. Petals split   . CHAILLETIACEÆ.
    Petals undivided   . BURSERACEÆ.

81. Carpels disunited   .   . 82
    Carpels consolidated   .   . 93
82. Anthers with recurved valves
                          BERBERACEÆ.
    Anthers with straight valves . 83
83. Fruit leguminous   .   . 84
    Fruit not leguminous   .   . 85
84. Radicle next hilum   . FABACEÆ.
    Radicle remote from hilum
                          CONNARACEÆ.
85. Leaves dotted   . AMYRIDACEÆ.
    Leaves not dotted   .   . 86
86. Carpels solitary . ANACARDIACEÆ.
    Carpels several .   .   . 87
87. Carpels with hypog. scales   . 88
    Carpels without ditto .   . 89
88. Hypogynous scales simple
                          CRASSULACEÆ.
    Hypogynous scales double
                          FRANCOACEÆ.
89. Herbaceous plants RANUNCULACEÆ.
    Trees or shrubs   .   . 90
90. Flowers unisexual MENISPERMACEÆ.
    Flowers hermaphrodite   . 91
91. Stamens perigynous
                          CALYCANTHACEÆ.
    Stamens hypogynous .   . 92
92. Stamens indefinite . ANONACEÆ.
    Stamens definite . CORIARIACEÆ.
93. Placenta dissepimental
                          NYMPHÆACEÆ.
    Placenta parietal   .   . 94
    Placenta axile .   .   . 99
94. Stamens tetradynamous
                          BRASSICACEÆ.
    Stamens not tetradynamous . 95
95. Hypogynous disk large   . 96
    Hypogynous disk small or 0   . 97
96. Stamens indefinite CAPPARIDACEÆ.
    Stamens definite . RESEDACEÆ.
97. Sepals tubular . FRANKENIACEÆ.
    Sepals distinct   .   . 98
98. Sepals 2-3   . PAPAVERACEÆ.
    Sepals 5   . TURNERACEÆ.
99. Brown parasites MONOTROPACEÆ.
    Green rooting plants .   . 100
100. Styles distinct .   .   . 101
     Styles consolidated   .   . 105
101. Stam. polyadelphous HYPERICACEÆ.
     Stamens free .   .   . 102

102. Carpels with an hypogynous
scale . . CRASSULACEÆ.
Carpels without do. . . 103
103. Carpels two divaricating
SAXIFRAGACEÆ.
Carpels parallel . . 104
104. Stigmas capitate . LINACEÆ.
Stigmas simple CARYOPHYLLACEÆ.
105. Stamens monadelphous . 106
Stamens free . . . 107
106. Seeds wingless . MELIACEÆ.
Seeds winged . CEDRELACEÆ.
107. Sepals 2 . . PORTULACACEÆ.
Sepals more than 2 . . 108
108. Anthers opening by pores
ERICACEÆ.
Anthers opening by slits . 109
109. Leaves dotted . . . 110
Leaves not dotted . . 112
110. Fruit succulent . AURANTIACEÆ.
Fruit capsular . . . 111
111. Flowers hermaphrodite RUTACEÆ.
Flowers polygamous
XANTHOXYLACEÆ.
112. Flowers irregular BALSAMINACEÆ.
Flowers regular . . 113
113. Stamens arising from scales
SIMARUBACEÆ.
Stamens not do. . . 114
114. Calyx valvate . . . 115
Calyx imbricated . . 116
115. Stam. opposite petals RHAMNACEÆ.
Stamens more numerous than
petals . . LYTHRACEÆ.
116. Flowers unisexual EMPETRACEÆ.
Flowers hermaphrodite . 117
117. Stamens hypogynous . . 118
Stamens perigynous CELASTRACEÆ.
118. Seeds comose . TAMARICACEÆ.
Seeds naked . PITTOSPORACEÆ.
119. Calyx none . . . 120
Calyx present . . . 128
120. Leaves with stipules . . 121
Leaves without stipules . 126
121. Ovules numerous . SALICACEÆ.
Ovules few . . . 122
122. Flowers hermaphrodite . 123
Flowers unisexual . . 124
123. Stam. unilateral CHLORANTHACEÆ.
Stamens whorled . SAURURACEÆ.

124. Carpels triple . EUPHORBIACEÆ.
Carpels single . . . 125
125. Ovule erect . . MYRICACEÆ.
Ovule pendulous . PLATANACEÆ.
126. Flowers hermaphrodite
PIPERACEÆ.
Flowers unisexual . . 127
127. Carpels single . MYRICACEÆ.
Carpels double . CALLITRICHACEÆ.
128. Ovary inferior . . . 129
Ovary superior . . 143
129. Leaves with stipules . . 130
Leaves without stipules . 132
130. Flowers hermaphrodite
ARISTOLOCHIACEÆ.
Flowers unisexual . . 131
131. Flowers amentaceous CORYLACEÆ.
Flowers not do. . BEGONIACEÆ.
132. Flowers unisexual . . 133
Flowers herm. or polyg. . 136
133. Climbing tendrilled herbs
CUCURBITACEÆ.
Trees or shrubs . . 134
134. Leaves compound . JUGLANDACEÆ.
Leaves simple . . . 135
135. Leaves opposite . GARRYACEÆ.
Leaves alternate . MYRICACEÆ.
136. Leaves dotted . MYRTACEÆ.
Leaves not dotted . . 137
137. Ovary 1-celled . . . 138
Ovary 2-6-celled . . . 142
138. Parasites on branches
LORANTHACEÆ.
Terrestrial . . . 139
139. Flowers $\frac{2}{1}$/ . . ONAGRACEÆ.
Flowers not $\frac{2}{1}$/ . . 140
140. Calyx valvate . SANTALACEÆ.
Calyx not valvate . . 141
141. Embryo straight COMBRETACEÆ.
Embryo curved CHENOPODIACEÆ.
142. Flowers $\frac{2}{1}$/ . . ONAGRACEÆ.
Flowers $\frac{3}{1}$/ . ARISTOLOCHIACEÆ.
143. Leaves with stipules . . 144
Leaves without stipules . 160
144. Flowers hermaphrodite . 145
Flowers unisexual . . 147
145. Sepals 2 . PORTULACACEÆ.
Sepals more than 2 . . 146
146. Carpels more than 1 consolidated 147
Carpels solitary . . 153

82 SYSTEMATICAL BOTANY.

147. Stamens hypogynous . . 148
Stamens perigynous . . 150
148. Fruit beaked . GERANIACEÆ.
Fruit not beaked . 149
149. Calyx biglandular imbricated
MALPIGHIACEÆ.
Calyx eglandular valvate TILIACEÆ.
150. Placenta parietal PASSIFLORACEÆ.
Placenta axile . . . 151
151. Leaves opposite . CUNONIACEÆ.
Leaves alternate . . 152
152. Calyx valvate . RHAMNACEÆ.
Calyx imbricate . ULMACEÆ.
153. Calyx membranous ILLECEBRACEÆ.
Calyx firm and herbaceous . 154
154. Styles from the base of ovary
CHRYSOBALANACEÆ.
Styles terminal . 155
155. Fruit leguminous . FABACEÆ.
Fruit not leguminous . . 156
156. Stipules ochreate POLYGONACEÆ.
Stipules simple or 0 . . 157
157. Styles simple . . ROSACEÆ
Styles triple . PETIVERIACEÆ.
158. Carpels solitary . URTICACEÆ.
Carpels more than one . . 159
159. Flowers amentaceous BETULACEÆ.
Flowers not amentaceous
EUPHORBIACEÆ.
160. Flowers hermaphrodite . 161
Flowers unisexual . . 183
161. Sepals 2 . . PORTULACACEÆ.
Sepals more than 2 . . 162
162. Carpels more than 2, consolidated 163
Carpels solitary or disjoined . 172
163. Placenta parietal PAPAVERACEÆ.
Placenta axile . . . 164
164. Ovules few . . . 165
Ovules many . . . 168
165. Leaves dotted . . RUTACEÆ.
Leaves not dotted . . 166
166. Leaves compound . OLEACEÆ.
Leaves simple . . . 167
167. Carpels numerous, separable
PHYTOLACCACEÆ.
Carpels few, inseparable
CELASTRACEÆ.
168. Carpels 2, divaricating
SAXIFRAGACEÆ.
Carpels not divaricating . 170

170. Stamens hypogynous
CARYOPHYLLACEÆ.
Stamens perigynous . . 171
171. Fruit one-celled . PRIMULACEÆ.
Fruit many-celled . LYTHRACEÆ.
172. Carpels several RANUNCULACEÆ.
Carpels solitary . . 173
173. Anther valves recurved LAURACEÆ.
Anther valves straight . . 174
174. Fruit a legume . . FABACEÆ.
Fruit not do. . . . 175
175. Leaves dotted . AMYRIDACEÆ.
Leaves not dotted . . 176
176. Stamens within the points of
sepals . . . PROTEACEÆ.
Stamens not do. . . 177
177. Calyx hardened . . 178
Calyx tube membranous . 179
178. Cal. all hardened SCLERANTHACEÆ.
Base only of cal. hardened
NYCTAGINACEÆ.
179. Fruit triangular POLYGONACEÆ.
Fruit rounded . . 180
180. Leaves lepidote ELÆAGNACEÆ.
Leaves not lepidote . . 181
181. Calyx tubular . THYMELACEÆ.
Calyx tubeless . . . 182
182. Calyx dry and coloured
AMARANTACEÆ.
Calyx herbaceous CHENOPODIACEÆ.
183. Carpels solitary or distinct . 184
Carpels consolidated . . 185
184. Calyx tubular, carpel solitary
MYRISTICACEÆ.
Calyx open, carpels several
MENISPERMACEÆ.
185. Leaves dotted . XANTHOXYLACEÆ.
Leaves not dotted EUPHORBIACEÆ.
186. Ovary superior . . 187
Ovary inferior . . 226
187. Flowers regular . . 188
Flowers irregular . . 218
188. Ovary deeply split . . 189
Ovary not split . . 192
189. Leaves dotted . . RUTACEÆ.
Leaves not dotted . . 190
190. Inflorescence gyrate BORAGINACEÆ.
Inflorescence straight . . 191
191. Æstivation plaited . NOLANACEÆ.
Æstivation flat STACKHOUSIACEÆ.

192. Carpels 4, 5, or more . . 193
     Carpels three . . . 204
     Carpels two . . . 206
     Carpels single . . . 216
193. Stamens opposite petals . 194
     Stamens alternate with petals . 195
194. Shrubs or trees . MYRSINACEÆ.
     Herbs . . PRIMULACEÆ.
195. Anthers opening by pores . 196
     Anthers opening by slits . 198
196. Anthers one-celled EPACRIDACEÆ.
     Anthers two-celled . . 197
197. Shrubs. Seeds wingless
                           ERICACEÆ.
     Herbs. Seeds winged PYROLACEÆ.
198. Brown parasites MONOTROPACEÆ.
     Rooting plants . . 199
199. Seeds numerous . CRASSULACEÆ.
     Seeds few or solitary . . 200
200. Carpels distinct . ANONACEÆ.
     Carpels consolidated . . 201
201. Ovules erect . . . 202
     Ovules pendulous . . 203
202. Æstivation imbricated SAPOTACEÆ.
     Æstivation plaited CONVOLVULACEÆ.
203. Stamens twice as many as petals
                           EBENACEÆ.
     Stamens equal to petals
                       AQUIFOLIACEÆ.
204. Inflorescence gyrate HYDROLEACEÆ.
     Inflorescence straight . . 205
205. Calyx imbricated CONVOLVULACEÆ.
     Calyx tubular . POLEMONIACEÆ.
206. Stamens 2 . . . 207
     Stamens more than 2 . . 208
207. Corolla valvate . OLEACEÆ.
     Corolla imbricate . JASMINACEÆ.
208. Inflorescence gyrate . . 209
     Inflorescence straight . . 210
209. Fruit two-celled . CORDIACEÆ.
     Fruit one-celled HYDROPHYLLACEÆ.
210. Corolla valvate . CESTRACEÆ.
     Corolla imbricate . . 211
211. Anthers united to stigma
                     ASCLEPIADACEÆ.
     Anthers free . . . 212
212. Corolla contorted . APOCYNACEÆ.
     Corolla imbricated or plaited . 213
213. Calyx broken-whorled . . 214
     Calyx imbricated . . 215

214. Leafless twiners . CUSCUTACEÆ.
     Leafy plants . CONVOLVULACEÆ.
215. Placentæ parietal . GENTIANACEÆ.
     Placentæ axile . SOLANACEÆ.
216. Stigma with an indusium
                       BRUNONIACEÆ.
     Stigma without an indusium . 217
217. Style single . PLANTAGINACEÆ.
     Styles 5 . PLUMBAGINACEÆ.
218. Ovary 4-lobed . LAMIACEÆ.
     Ovary undivided . . 219
219. Carpel solitary . GLOBULARIACEÆ.
     Carpels two . . . 220
220. Fruit nut-like . VERBENACEÆ.
     Fruit capsular or succulent . 221
221. Placenta parietal OROBANCHACEÆ.
     Placenta free, central
                       LENTIBULACEÆ.
     Placenta axile . . 222
222. Seeds winged . BIGNONIACEÆ.
     Seeds wingless . . 223
223. Placentæ double CYRTANDRACEÆ.
     Placentæ simple . . 224
224. Ovary partly inferior GESNERACEÆ.
     Ovary quite superior . . 225
225. Calyx broken-whorled
                       ACANTHACEÆ.
     Calyx tubular or imbricated
                   SCROPHULARIACEÆ.
226. Carpel solitary . . 227
     Carpels more than one . . 229
227. Anthers syngenesious ASTERACEÆ.
     Anthers free . . . 228
228. Flowers in heads . DIPSACEÆ.
     Flowers loose . VALERIANACEÆ.
229. Anthers syngenesious LOBELIACEÆ.
     Anthers free . . . 230
230. Anthers opening by pores
                       VACCINACEÆ.
     Anthers opening by slits . 231
231. Stipules between the leaves
                       CINCHONACEÆ.
     Stipules none . . 232
232. Stigma with an indusium
                       GOODENIACEÆ.
     Stigma naked . . . 233
233. Seeds indefinite . . 234
     Seeds few in number . . 235
234. Stamens free . CAMPANULACEÆ.
     Stamens consolidated STYLIDIACEÆ.

235. Leaves alternate . EBENACEÆ.
Leaves opposite . . 236
236. Fruit didymous . GALIACEÆ.
Fruit not didymous CAPRIFOLIACEÆ.
237. Stem cylindrical, unbranched
CYCADACEÆ.
Stem conical, branched . . 238
238. Fruit solitary . . TAXACEÆ.
Fruit in cones . . PINACEÆ.
239. Ovary inferior . . . 240
Ovary superior . . 246
240. Flowers gynandrous ORCHIDACEÆ.
Flowers not gynandrous . 241
241. Anther 1 . . . 242
Anthers more than one . 243
242. Anther one-celled . MARANTACEÆ.
Anther two-celled ZINGIBERACEÆ.
243. Veins of leaves diverging MUSACEÆ.
Veins of leaves straight . 244
Veins of leaves netted DIOSCOREACEÆ.
244. Stamens 3 . . IRIDACEÆ.
Stamens 6 . . . 245
Stamens more than 6
HYDROCHARACEÆ.
245. Sepals petaloid . AMARYLLIDACEÆ.
Sepals herbaceous BROMELIACEÆ.
246. Flowers glumaceous . . 247
Flowers regular . . 248
247. Sheaths of leaves slit GRAMINACEÆ.
Sheaths of leaves closed
CYPERACEÆ.
248. Leaves netted . . SMILACEÆ.
Leaves straight-veined . 249
249. Carpels disunited . . 250
Carpels consolidated . . 259
250. Placentæ dissepimental BUTOMACEÆ.
Placentæ axile . . . 251
251. Flowers imperfect . . 252
Flowers perfect . . 257
252. Flowers on a spadix . . 253
Flowers scattered . . 255
253. Fruit drupaceous . PANDANACEÆ.
Fruit berried . . ARACEÆ.
Fruit dry . . . 254

254. Stamens very short . ACORACEÆ.
Stamens long, weak . TYPHACEÆ.
255. Floating . . . 256
Terrestrial . JUNCAGINACEÆ.
256. Ovules pendulous . NAIADACEÆ.
Ovules erect . . PISTIACEÆ.
257. Anthers turned outwards
MELANTHACEÆ.
Anthers turned inwards . 258
258. Stems herbaceous . ALISMACEÆ.
Stems woody . PALMACEÆ.
259. Flowers semipetaloid
COMMELINACEÆ.
Flowers hexapetaloid . . 260
260. Flowers coloured . LILIACEÆ.
Flowers scarious . JUNCACEÆ.
261. Axis distinct, leafy . . 262
Axis distinct, leafless CHARACEÆ.
Axis confused . . . 268
262. Sporangia upon leaves FILICALES.
Sporangia arising from the stem 263
263. Sporangia involucrate . . 264
Sporangia naked . . 265
264. Involucres uniform
MARSILEACEÆ.
Involucres 2-formed SALVINIACEÆ.
265. Sporangia axillary, sessile, 2-
valved . LYCOPODIACEÆ.
Sporangia stalked . . 266
266. Sporangia valveless . BRYACEÆ.
Sporangia valvate . . 267
267. An operculum . ANDRÆACEÆ.
No operculum JUNGERMANNIACEÆ.
268. Stomates . . . 269
No stomates . . . 270
269. Sporangia valvate, operculate
JUNGERMANNIACEÆ.
Sporangia valveless, without an
operculum . MARCHANTIACEÆ.
270. Submersed . . ALGACEÆ.
Aerial 271
271. Thallus superficial LICHENACEÆ.
Thallus buried . FUNGACEÆ.

III.—THE NATURAL SYSTEM.

627. THE true Natural System, whenever it shall be discovered, will represent the species, genera, orders, alliances, groups, subclasses, and classes of plants, or whatever other divisions may be admitted into it, so arranged that each plant shall stand next those to which it is more nearly allied in structure than to any others.

628. But the skill of man has not yet attained this end; no system answering to this description has been devised, nor does there appear any probability that it will be discovered till our knowledge of plants is much more advanced.

629. All so-called natural systems are, to the present day, partly artificial and partly natural. The lower and higher divisions in them are natural, the intermediate divisions are artificial. In other words, the stones of the edifice are hewed and squared, and the general plan is drawn out, but no builder has yet been found with skill to put them together, so as to form a consistent whole.

630. But although in theory no system that can properly be called natural has yet been devised, yet for practical purposes many answer to the name, and fulfil the principal conditions required of them.

631. The genera and natural orders can alone be considered as agreed upon by botanists, the other divisions are unsettled; and this is the reason why the natural orders seldom follow in the same manner in the arrangements of two different botanists.

632. There is no such thing as an arrangement which shall express the natural relations of plants in a consecutive series.

633. It seems to be generally admitted by those who have turned their attention to the consideration of the manner in which organized beings are related to each other, that each species is allied to many others in different degrees, and that such relationship is best expressed by rays (the affinities)

proceeding from a common centre (the species). In like manner, in studying the mutual relationship of the several parts of the vegetable kingdom, the same form of distribution constantly forces itself upon the mind; genera and orders being found to be apparently the centre of spheres, whose surface is only defined by the points where the last traces of affinity disappear.

634. But although the mind may conceive such a distribution of organized beings, it is impossible that it should be so presented to the eye, and all attempts at effecting that object have failed. If in describing the surface of a sphere we are compelled to travel in various directions, continually returning back to the point from which we started, and if in presenting it to the eye at one glance we are compelled to project it upon a plane, the effect of which is to separate to the greatest distance some objects which naturally touch each other, how much more impossible must it be to follow the juxtaposition of matter in treating of the solid contents of a sphere.

635. The fundamental principle of systematic botany is, that those plants should be stationed in company with each other which have the greatest degree of affinity, and that those should be placed most remotely which have the smallest degree of affinity.

636. Affinity is an accordance in all essential characters.

637. From this is distinguished analogy, which is a conformity in one or two characters only.

638. What we call the characters of plants are merely the signs by which we judge of affinity, and all the groups into which plants are thrown are in one sense artificial, inasmuch as nature recognises no such groups.

639. Nevertheless, consisting in all cases of species very closely allied in nature, they are in another sense natural.

640. But as the classes, subclasses, groups, alliances, natural orders, and genera of botanists have no real existence in nature, it follows that they have no fixed limits, and consequently that it is impossible to define them.

641. They are to be considered as nothing more than the expression of particular *tendencies* (nixus), on the part of the

plants they comprehend, to assume a particular mode of developement.

642. Their characters are therefore nothing more than a declaration of their prevailing tendencies, and are liable to numerous exceptions.

This liability, it must be remarked, exists as much in all artificial schemes as in the natural system itself.

643. If a system is ever to be devised which shall by common consent be admitted to be natural in all its parts, as far as human means can make it so, this will be brought about by settling the relative value of the characters by which plants are limited, and by introducing uniformity and consistency into the distinctions of the groups, whether inferior, superior, or intermediate.

Up to the present time, no attempt at settling these points has been successful, and consequently the characters employed in defining the limits of groups, of all denominations except the highest, are arbitrary and inconsistent.

644. The following propositions seem incontrovertible :—
1. Nothing that is constant can be regarded as unimportant.
2. Every thing constant must be dependent upon or connected with some essential function. Therefore all constant characters, of whatever nature, require to be taken into account in classifying plants according to their natural affinities.

Of this nature are the internal structure of stems and leaves, the anatomical condition of tissue, the organization of the anther, pollen, and female apparatus, and the interior of the seed.

645. On the other hand, whatever points of structure are variable in the same species, or in species nearly allied to each other, or in neighbouring genera, are unessential to the vital functions, and should be set aside, or be regarded as of comparative unimportance.

Hence the badness of the Monopetalous, Polypetalous, and Apetalous divisions of Jussieu, depending upon the presence or absence, and union or disunion, of petals. The genus Fuchsia, for example, has petals highly developed ; but in F. excorticata they are absent, and yet the plant differs no otherwise from the rest of the genus : the same is true of species of Rhamnus. Again, the Rue has the petals separate ; and Correa, very nearly allied to it, has them combined.

646. Those peculiarities of structure which are connected with the manner in which a plant is developed are *physiological*.

647. Those peculiarities of structure which are connected with the manner in which parts are arranged are *structural*.

648. Physiological characters are of two kinds; 1, those which are connected with the *mode of growth* (or *organs of vegetation*), and, 2, those which regulate *reproduction* (or *organs of fructification*).

649. Physiological characters are of greater importance in regulating the natural classification of plants than structural.

650. All modifications of either are respectively important, in proportion to their connection with the phenomena of life.

651. If we allow ourselves to be steadily guided by these considerations, we shall find that the internal or anatomical structure of the axis, and of the foliage, is of more importance than any other character.

> Because these are the circumstances which essentially regulate the functions of growth, and the very existence of an individual.

652. That next in order is the internal structure of the seed, by which the species must be multiplied.

> Thus the presence of an embryo, or its absence, the first indicating a true seed (531), the latter a spore (590), are most essential circumstances to consider. And so also the existence of albumen in abundance round the embryo, or its absence, must be regarded as a physiological character of the highest value : because, in the former case, the embryo demands a special external provision for its early nutriment, as in oviparous animals ; while, in the latter case, the embryo is capable of developing by means of the powers resident in itself, and unassisted, as in viviparous animals.

653. Next to this must be taken the structure of the organs of fructification, by whose united action the seed is engendered; for without some certain, uniform, and invariable action on their part, the race of a plant must become extinct.

> Thus we find that the structure of the anthers, placentæ, and ovules, are more uniform than that of the parts surrounding them ; while their numbers are variable ; and the condition of the filament, which appears of so little importance in a physiological point of view, is also inconstant. So also the texture and surface and form of the pericarp, which acts as a mere covering to the seeds, is not to be regarded in these inquiries, and, in fact, differs from genus to genus ; as, for instance, between Pyrus and Stranvæsia, or Rubus and Spiræa, in the truly natural Rosaceous order.

654. On the other hand, of the floral envelopes (322), the number, form, and condition, the presence or absence, the regularity or irregularity, seem to be unconnected with functions of a high order, and to be designed rather for the decoration of plants, or for the purpose of giving variety to the aspect of the vegetable world; they are consequently of low and doubtful value, except for specific distinctions.

There seems, indeed, reason to expect that every natural order will, sooner or later, be found to contain within itself all the variations above alluded to. Even in the cases of regularity and irregularity we already know this to be so ; witness Veronica and Scoparia in Scrophulariaceæ, and Hyoscyamus in Solanaceæ, Delphinium in Ranunculaceæ, and Pelargonium in Geraniaceæ.

655. The consolidation of the parts of fructification is a circumstance but little attended to in a general point of view, except in respect to the corolla; but as it seems to indicate either the greatest change that the parts can undergo, or, where it occurs between important and unimportant organs, that in such cases the latter are essential to the former, it probably deserves to be regarded with great attention.

For instance, the presence or absence of the corolla is often a point of little moment, and is, we know, a very fluctuating circumstance. This is especially true of those natural orders in which the stamens and petals are separated ; as in Rosaceæ, Rhamnaceæ, Onagraceæ, &c. On the other hand, when the stamens, which are indispensable organs, adhere to the petals, the latter are more constantly present, as in Scrophulariaceæ, Acanthaceæ, Solanaceæ, &c.

656. If consolidation is, on the one hand, to be regarded as a character of high importance, so must disunion also be so considered on the other.

This is indicated by those natural orders of plants, which, like the Rosaceous, the Ranunculaceous, and the Magnoliaceous, are called apocarpous.

657. If we descend lower than those points, we find it extremely difficult, when we enter into details, to comprehend what gives some of the subordinate peculiarities of plants the value we assign to them. No fixed rule has yet been discovered for judging of this ; and the employment of secondary characters is in a great degree arbitrary.

IV.—THE NATURAL SYSTEM OF DE CANDOLLE.

MANY natural systems have been proposed by different botanists. Ray, Linnæus, Jussieu, De Candolle, Bartling, Reichenbach, Schultz, Endlicher, myself, and many others, have each had their own system; and, perhaps, the best character that can be given of them is, that while they are all far from the truth, each has some merits which the others want.

The system of De Candolle, however, having been taken as the basis of the most perfect enumeration of plants that has ever been made, has so great a reputation, that for the convenience of students it most requires explanation. And it seems the more deserving of illustration, because the University of London have declared that their examinations shall be conducted with reference to it.

It will not be necessary to introduce into an illustration of this system every natural order; for many are imperfectly known, and only interest the botanist when he extends his inquiries into the minutiæ of the science. All, however, of importance, will be found in the succeeding pages.

Plants are either furnished with visible flowers, or they are multiplied in some other way. Hence the two great divisions, of FLOWERING (*Phænogamous* or *Phanerogamous*), and FLOWERLESS (*Cryptogamous*).

Flowering plants are either EXOGENS (95) or ENDOGENS (95), with which *Dicotyledons* (573) and *Monocotyledons* (572) respectively correspond.

Flowerless plants are either ÆTHEOGAMOUS (*Semivascular*), that is, furnished with stomates and vascular tissue; or they are AMPHIGAMOUS (*Cellular*), that is, destitute of stomates and entirely cellular.

Hence arise four CLASSES.

I. FLOWERING PLANTS.

Class 1. Exogens *or* Dicotyledons.
Class 2. Endogens *or* Monocotyledons.

II. FLOWERLESS PLANTS.

Class 3. Ætheogamous *or* Semivascular.
Class 4. Amphigamous *or* Cellular.

CLASS I. EXOGENÆ.

This is the largest class in the vegetable kingdom, comprehending more species than all the others put together. The subclasses are the following :

1. *Thalamifloræ.* A calyx and corolla. Petals distinct; Stamens hypogynous.

2.* *Calycifloræ.* A calyx and corolla. Petals distinct; Stamens perigynous.

3. *Corollifloræ.* A calyx and corolla. Petals united, bearing the stamens.

4. *Monochlamydeæ.* A calyx only, or none.

SUBCLASS I. THALAMIFLORÆ.

Order 1.—*Ranunculaceæ.* Herbs or shrubs, occasionally climbing. Leaves with the petiole generally dilated, and the blade very often palmate or digitate. Sepals 3-6, usually deciduous. Petals 3-15, or none. Stamens indefinite; anthers adnate. Carpels numerous, or united into a single pistil. Seeds either erect or pendulous.

USES.—Generally acrid, bitter, narcotic plants, with vesicating leaves, as Aconite, Stavesacre, Crowfoot. Some however have the bitter principle predominant and the acridity slight, as Hydrastis canadensis, Coptis, Xanthorhiza, which are tonics.

---

* These are not exactly the characters given by De Candolle, who includes all monopetalous orders with an inferior ovary in Calycifloræ, and limits Corollifloræ to the hypogynous monopetalous orders. But it seems to me more easy in practice to regard Corollifloræ as equivalent to the Monopetalæ of Jussieu, while Thalamifloræ and Calycifloræ correspond to the Polypetalæ of that author, and Monochlamydeæ to his Apetalæ ; and in a series so very artificial as this, we may be permitted, I think, to consult convenience.

This order divides into two principal sections.

I. Flowers regular.   Typical Genera.—Ranunculus, Clematis, Adonis.

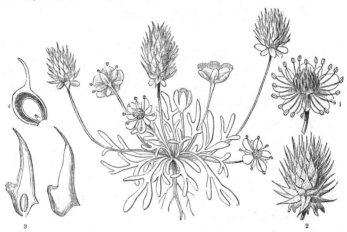

Ceratocephalus orthoceras.   1. Flower.   2. Ripe fruit.   3. Ovaries of Ranunculus Krapfia.   4. Section of carpel and seed of the same.

II. Flowers irregular.   Typical Genera.— Delphinium, Aconitum.

Delphinium tricorne.   1. Petals and stamens.   2. Carpels.   3. A branch of ripe fruit.

2.—*Anonaceæ.*   Trees or shrubs generally tropical.   Leaves without stipules.   Flowers axillary, large, and dull-coloured.

Sepals 3-4. Petals 6, coriaceous, with a valvular æstivation.
Stamens indefinite; anthers adnate; filaments angular.
Ovaries numerous. Fruit succulent or dry, with the carpels 1 or many-seeded, separate or consolidated. Embryo
minute. Albumen ruminate.

Uses.—Aromatic and fragrant in most cases. The fruits
of some are succulent and eatable, as the Custard Apple,
Anona squamosa, and the Cherimoyer, Anona Cherimolia;
those of others are hard, dry, and often jointed, as Habzelia
aromatica, the Piper Æthiopicum of the shops, and are used as
peppers. Some species are employed as febrifuges.

Typical Genera.—Anona, Uvaria.

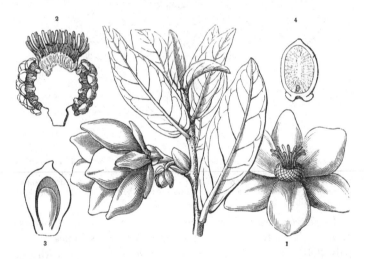

Anona furfuracea. 1. An expanded flower. 2. A vertical section of the andrœceum and gynœceum, which latter forms a central and terminal tuft. 3. A vertical
section of a carpel. 4. A vertical section of a ripe seed, showing the embryo and
ruminated albumen.

3.—*Menispermaceæ.* Shrubs with a sarmentaceous habit.
Leaves alternate. Flowers small. Flowers unisexual, usually
very small. Sepals in one or several rows. Stamens monadelphous or distinct. Anthers turned outwards. Ovaries
numerous, each with one style, sometimes soldered together
into a many-celled body, which is occasionally, in consequence of abortion, 1-celled. Drupes berried, 1-seeded
embryo curved, lying in albumen; radicle superior.

Uses.—Roots of many bitter and tonic, as Cocculus palmatus, which yields the Calumba root; of others also diuretic, as Cissampelos Pareira, and Cocculus Bakis, the latter a remedy used by the negroes of Senegal against intermittents. In the seeds a poison is formed, which in Anamirta Cocculus, the Cocculus Indicus of the shops, becomes extremely dangerous.

Typical Genera.—Menispermum, Cocculus.

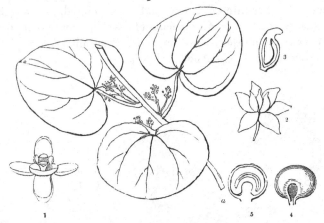

Cissampelos Pareira. 1. A male flower. 2. A female flower. 3. The vertical section of an ovary, which gradually curves the apex downwards, till, when it becomes the drupe 4, it acquires a horseshoe form. 5. A vertical section of a drupe, showing the embryo and albumen; *a.* is the true apex of the fruit, brought to the base as just described.

4.—*Berberaceæ.* Shrubs or herbaceous perennial plants. Leaves alternate, compound, usually without stipules. Sepals 3-4-6, in a double row. Petals sometimes with an appendage at the base. Stamens equal in number to the petals, and opposite to them; anthers opening elastically with a valve from the bottom to the top. Ovary solitary, 1-celled. Seeds attached to the bottom of the cell, 1, 2, or 3; albumen between fleshy and corneous.

Uses.—Bark astringent, and in Berberis yielding a yellow die. Fruit of Berberis acid; tubers of Bongardia eatable.

Typical Genera.—Berberis, Epimedium.

5.—*Nymphæaceæ.* Herbs with peltate or cordate fleshy leaves, growing in quiet water. Sepals and petals imbricated, passing gradually into each other. Stamens numerous, inserted above the petals into the disk; filaments petaloid; disk large, fleshy. Fruit many-celled. Seeds very numerous, attached to spongy dissepiments. Embryo on the outside of the base of the albumen, in a bag.

Uses.—Of little moment. Euryale seeds are eaten. Rhizomata slightly astringent and sedative.

Typical Genera.—Nymphæa, Nuphar.

6.—*Nelumbiaceæ.* Herbs with peltate, floating leaves. Sepals 4 or 5. Petals numerous. Stamens numerous. Disk fleshy, enclosing in hollows of its substance the monospermous ovaries. Nuts numerous, half buried in the disk.

Uses.—Nuts and creeping rhizomata eatable.

Typical Genus.—Nelumbium.

7.—*Dilleniaceæ.* Trees, shrubs, or under-shrubs, rarely herbaceous, leaves without stipules. Flowers often yellow. Sepals 5; 2 exterior, 3 interior. Petals 5. Stamens indefinite. Ovaries definite. Carpels baccate or 2-valved. Seeds surrounded by a pulpy aril. Embryo in solid albumen.

Uses.—Generally astringent. The leaves of many species are covered with asperities, which render them useful mechanically as polishing substances. Nothing deleterious known among them. Flowers occasionally intolerably fœtid.

Typical Genera.—Dillenia, Tetracera, Hibbertia.

8.—*Magnoliaceæ.* Trees or shrubs with convolute stipules. Flowers large, solitary. Sepals 3-6. Petals 3-27, imbricated. Stamens indefinite. Carpels numerous, distinct or consolidated.

Uses.—Bark tonic and febrifugal; that of the root of Magnolia glauca and Liriodendron in great repute in North America. Flowers often very fragrant.

Typical Genera.—Magnolia, Liriodendron.

9.—*Winteraceæ.* Shrubs or small trees. Woody tissue glandular. Leaves alternate, dotted, with convolute deciduous stipules. Flowers often brown. Sepals 2-6. Petals 2-30, imbricated. Stamens hypogynous. Ovaries 1-celled with suspended or erect ovules. Fruit consisting of a single row of carpels. Seeds with or without aril.

Uses.—Aromatic stimulants. An Illicium yields the Star Anise, and Drimys Winteri, the Winter's Bark, of the shops.

Typical Genera.—Illicium, Tasmannia.

10.—*Fumarieæ.* Herbaceous plants with brittle stems and a watery juice. Sepals 2. Petals 4; parallel; the outer one, or both saccate at the base. Stamens 6, in 2 parcels.

Uses.—Unimportant. Species slightly diaphoretic.

Typical Genera.—Fumaria, Corydalis.

Fumaria officinalis. 1. A flower seen from below. 2. The same from the side. 3. The pistil, stamens, and a portion of the bagged upper sepal. 4. A parcel of anthers. 5. The fruit.

11.—*Sarraceniaceæ.* Herbaceous perennial plants, living in bogs. Leaves with a hollow urn-shaped petiole. Scapes bearing one large flower. Sepals 5, imbricate. Petals 5, unguiculate, concave. Stamens indefinite, hypogynous. Ovary 5-celled ; stigma very large, umbrella-shaped, peltate. Capsule crowned by the stigma. Seeds very numerous, minute.

Uses.—Unknown. Petiole-like leaves remarkable.

Typical Genus.—Sarracenia.

12.—*Brassicaceæ* or *Cruciferæ*. Herbaceous plants; rarely under-shrubs. Leaves alternate. Flowers without bracts. Sepals 4, deciduous, cruciate. Petals 4, cruciate. Stamens 6, of which two are shorter (tetradynamous). Ovary superior, with parietal placentæ, meeting in the middle, and forming a spurious dissepiment. Fruit a silique or silicule. Seeds attached by a funiculus, generally pendulous. Embryo with the radicle folded upon the cotyledons.

A very large and difficult natural order, the subdivisions in which are now made to depend upon the structure of the embryo. They are the following :

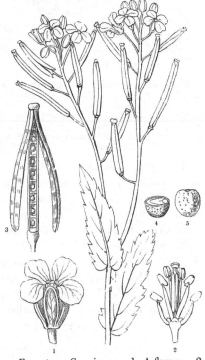

1. *Pleurorhizeæ*, when the embryo has the radicle applied to the edge of the cotyledons; *fig.* 17.

2. *Notorhizeæ*, when the embryo has the radicle applied to the back of the cotyledons; *fig.* 14.

3. *Orthoploceæ*, when the embryo has the radicle applied to the back of cotyledons which are hollowed out; *fig.* 12.

4. *Diplecolobeæ*, when the cotyledons are three times folded, and the radicle applied to their back; *fig.* 16.

Uses.—All the species harmless; some antiscor-

Erucastrum Canariense. 1. A flower. 2. The stamens. 3. The siliqua, with the valves separating from the replum. 4. A transverse section of a seed. 5. A perfect seed.

H

butic, all more or less pungent. Radishes, Turnips, Mustard, Cress, Cabbage and all its varieties, Rape, Charlock, are well-known plants of the order.

TYPICAL GENERA.—Brassica, Sinapis, Draba.

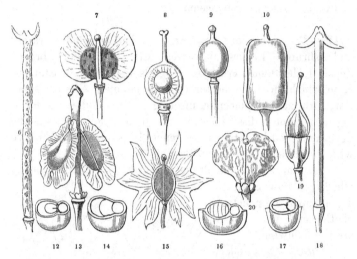

Fruits of various genera. 6. Siliqua of Mathiola livida. 7. Silicula of Thlaspi latifolium. 8. Silicula of Alyssum spathulatum. 9. Silicula of Schiverukia podolica. 10. Silicula of Farsetia. 12. Seed of Didesmus Ægyptius cut across. 13. Silicula of Menonvillea linearis. 14. Seed of Lepidium Africanum. 15. Silicula of Æthionema cristatum. 16. Seed of Heliophila crithmifolia. 17. Seed of Mathiola oxyceras. 18. Siliqua of Mathiola oxyceras. 19. Silicula of Didesmus Ægyptius. 20. Silicula of Senebiera serrata.

13.—*Papaveraceæ.* Herbaceous plants or shrubs with a milky juice. Leaves alternate, without stipules. Sepals 2. Petals either 3 or 4, or some multiple of that number. Stamens hypogynous, generally numerous. Fruit 1-celled, with parietal placentæ. Seeds numerous.

USES.—A narcotic milk pervades the species; that of Papaver somniferum becomes opium when inspissated. The roots of Meconopsis Nepalensis are a deadly poison. Sanguinaria Canadensis is emetic and purgative in large doses, stimulant and diaphoretic in smaller.

TYPICAL GENERA.—Papaver, Glaucium.

14.—*Capparidaceæ.* Herbaceous plants, shrubs, or trees, without true stipules. Leaves alternate. Sepals 4. Petals 4, cruciate. Stamens definite or indefinite. Disk hemispherical,

or elongated.   Ovary stalked.   Fruit 1-celled, most frequently with two polyspermous placentæ ; embryo incurved.

Uses.—A pungent principle exists in some, as the flower-
buds of Capparis spinosa, which are the Capers of shops, and
several Cleomes used as substitutes for mustard.   This
acridity is sometimes so much concentrated as to render the
species dangerous.   The root of Cratæva gynandra is said to
blister like Cantharides, and that of Cleome dodecandra is
used as a vermifuge.

Typical Genera.—Cleome, Capparis.

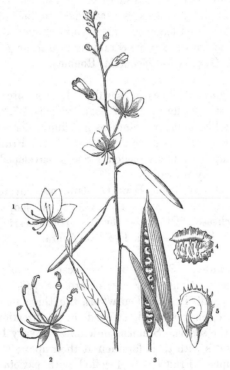

Physostemon lanceolatum. 1. A flower of the natural size. 2. The calyx, stamens, and ovary. 3. The ripe fruit, with one valve separating. 4. A seed. 5.
The same cut vertically, to show the incurved embryo.

15.—*Resedaceæ.* Herbaceous plants with alternate leaves,
small colourless flowers, and gland-like stipules. Calyx
many-parted. Petals lacerated, unequal. Disk large, 1-sided.
Stamens definite, inserted into the disk.   Ovary sessile,

3-lobed, 1-celled, many-seeded, with 3 parietal placentæ. Fruit opening at the apex. Embryo incurved.

Uses.—Mignionette, so well known for its fragrance, is Reseda odorata. R. luteola yields a yellow dye.

Typical Genera.—Reseda, Ochradenus.

16.—*Flacourtiaceæ.* Shrubs or trees. Leaves alternate, without stipules. Sepals from 4-7. Petals equal to them in number. Stamens occasionally changed into nectariferous scales. Ovary roundish; stigmas several, more or less distinct. Fruit 1-celled, capsular or fleshy, the centre filled with a thin pulp. Seeds few, attached to the lining of the fruit in a branched manner. Embryo in albumen.

Uses.—The fruit of some eatable and pleasant in India.

Typical Genera.—Flacourtia, Roumea.

17.—*Bixaceæ.* Trees or shrubs. Leaves alternate, with deciduous stipules and pellucid dots. Sepals 4-7, imbricated. Petals of a like number. Stamens indefinite, distinct. Ovary sessile; placentæ 4-7, parietal; styles 1-2-4. Fruit 1-celled, fleshy or capsular, many-seeded. Seeds enveloped in pulp. Albumen hardly present.

Uses.—The seeds of Bixa Orellana are covered with a pulp, which, when dry, is the Arnotta of shops, used for colouring cheese. Otherwise the properties uncertain.

Typical Genera.—Bixa, Prockia, Azara.

18.—*Cistaceæ.* Shrubs or herbaceous plants. Leaves usually entire, stipulate or exstipulate. Sepals 3 or 5, persistent, unequal, in a broken whorl, the three inner twisted. Petals 5, often crumpled, twisted in a direction contrary to that of the sepals. Stamens indefinite. Ovary 1- or many-celled; ovules with their foramen at their apex; style single; stigma simple. Fruit either 1-celled with parietal placentæ, or imperfectly 5- or 10-celled. Seeds indefinite. Embryo inverted, either spiral or curved, in the midst of mealy albumen. Radicle remote from the hilum.

Uses.—Unimportant. The balsamic Gum Ladanum is a spontaneous secretion from Cistus Creticus and others. Many are beautiful garden plants, with large delicate flowers.

Typical Genera.—Cistus, Helianthemum.

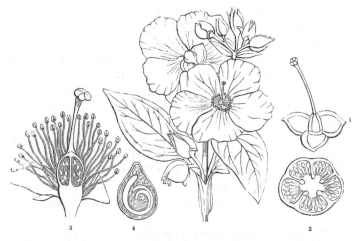

Cistus Berthelotianus. 1. A calyx and pistil, the petals and stamens having fallen off. 2. A cross section of the ovary. 3. A vertical section of ovary and calyx. 4. A seed cut through ; the pointed end being the true apex.

19.—*Droseraceæ.* Herbaceous plants, often covered with glands. Leaves with stipulary fringes and a circinate vernation. Peduncles circinate. Sepals imbricate. Petals 5, hypogynous. Stamens distinct, either equal in number to the petals, or 2, 3, or 4 times as many. Styles 3-5. Capsule of 3-5 valves. Embryo in fleshy or cartilaginous albumen.

Uses.—The herbage of some Droseras is acrid. The bulbs of others abound in a rich purple dye, and are filled with starch, which renders them eatable. It is probable that many species would prove of value to dyers.

Typical Genera.—Drosera, Dionæa.

20.—*Tamaricaceæ.* Shrubs or herbs, with rod-like branches. Leaves alternate, resembling scales. Calyx 4- or 5-parted, persistent. Petals withering. Stamens definite, distinct, or monadelphous. Stigmas 3. Capsule 3-valved, 1-celled, many-seeded. Seeds ascending, comose ; embryo straight.

Uses.—Ornamental bushes or trees. A sweet substance resembling Manna oozes out of the stem of Tamarix Gallica, in hot, dry countries. The bark is bitter, astringent, and tonic. A very astringent gall, employed in medicine and dyeing, in India, is yielded by some oriental species.

Typical Genera.—Tamarix, Myricaria.

21.—*Violaceæ.* Herbaceous plants, or shrubs, or trees. Leaves stipulate, with an involute vernation. Sepals 5, persistent, imbricate, Petals 5, regular or irregular, one sometimes spurred. Stamens definite in number; filaments dilated; connective elongated beyond the anthers. Ovary 1-celled, with 3 parietal placentæ; style with a hooded stigma. Capsule of 3 valves, bearing the placentæ in their axis. Embryo large, straight, in fleshy albumen.

USES.—Roots emetic. Those of the common Sweet Violet and other species have been employed medicinally. Ionidium Poaya yields one sort of Brazilian Ipecacuanha. Viola canina and some others have the power of removing some cutaneous affections, and have been employed as cosmetics.

TYPICAL GENERA.—Viola, Alsodeia.

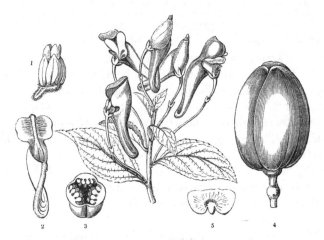

Corynostylis Hybanthus. 1. A set of stamens, each having the connective lengthened beyond the anther in the form of a scale. 2. A spurred petal. 3. A transverse section of an ovary, showing the three parietal placentæ. 4. A ripe fruit. 5. An embryo.

22.—*Polygalaceæ.* Shrubs or herbaceous plants. Leaves alternate, destitute of stipules. Pedicels with three bracts. Flowers unsymmetrical. Sepals 5, very irregular, often glumaceous. Petals consolidated, hypogynous, usually 3, of which 1 is anterior and larger than the rest. Stamens usually in a tube; anthers innate, 1-celled, and opening at their apex.

Ovary with 2 or 3 cells; ovules solitary, pendulous. Seeds pendulous, with a caruncula next the hilum; albumen abundant.

USES.—Leaves bitter, root milky. Polygala Senega, the Rattlesnake root, is stimulant, diaphoretic, emetic, and emmenagogue; it has been employed successfully in croup. Many other species have similar properties. Polygala Poaya is one of the Brazilian emetics. The bark of Monnina polystachya, a Peruvian plant, is detersive, and used as a substitute for soap.

TYPICAL GENERA.—Polygala, Muraltia, Mundia.

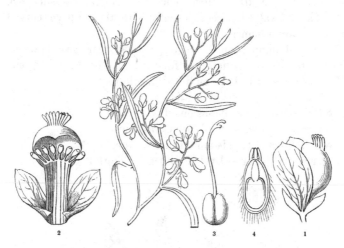

Polygala erioptera. 1. An entire flower seen from the side. 2. The same cut open to exhibit the stamens. 3. The pistil. 4. A section of a ripe seed; in the middle is the embryo; at the apex, which represents the real base, is seen a caruncula.

23.—*Frankeniaceæ.* Herbaceous plants or under-shrubs. Stems much branched. Leaves small, opposite, exstipulate, with a membranous sheathing base. Sepals 4-5, in a furrowed tube. Petals hypogynous, unguiculate, with appendages at the base of the limb. Stamens definite. Style 2- or 3-fid. Capsule 1-celled, enclosed in the calyx, 2- 3- or 4-valved, many-seeded. Seeds attached to the margins of the valves, very minute; embryo in the midst of albumen.

USES.—Unknown.

TYPICAL GENUS.—Frankenia.

24.—*Elatinaceæ.* Little weedy annuals. Leaves opposite, with stipules. Flowers minute. Sepals 3-5. Petals hypogynous. Stamens definite. Ovary 3-5-celled; styles 3-5; stigmas capitate. Fruit capsular. Seeds numerous, embryo straight, with but little albumen.

Uses.—Unknown.

Typical Genera.—Elatine, Bergia.

25.—*Caryophyllaceæ.* Herbaceous plants with opposite undivided exstipulate leaves, and tumid nodes. Sepals 4-5. Petals often slit. Stamens definite. Ovary usually many-seeded, with a free central placenta. Stigmas sessile, 2-5. Capsule 2-5-valved. Seeds usually with the embryo curved round mealy albumen.

Uses.—Unimportant. Some species bear gay flowers; more are weeds; a few are fragrant, as the Pink. Silene Virginica is said to have an anthelmintic root.

There are two sections of this order:

§1. *Alsineæ.* Sepals disjoined.

Typical Genera.—Stellaria, Cerastium.

§2. *Sileneæ.* Sepals united into a tube.

Typical Genera.—Lychnis, Silene, Dianthus.

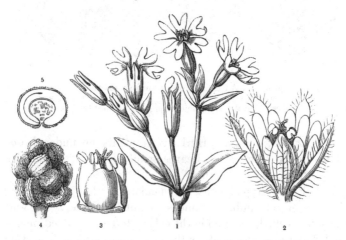

1. Lychnis diurna, § *Sileneæ.* 2. A flower of Stellaria media, § *Alsineæ.* 3. Its stamens and pistil. 4. Its placenta loaded with seeds. 5. A seed cut through vertically, to show the embryo curved round mealy albumen.

26. — *Malvaceæ.* Herbaceous plants, trees, or shrubs. Leaves alternate, stipulate, very often covered with stellate hairs. Flowers generally showy. Calyx with a valvate æstivation. Petals twisted. Stamens indefinite, monadelphous; anthers 1-celled, reniform. Ovary formed by the union of several carpels; styles the same number as the carpels. Fruit either capsular or baccate; albumen in small but variable quantity; embryo curved, with twisted and doubled cotyledons.

USES.—Mucilaginous; as Marsh Mallow and common Mallow. The unripe fruit of Hibiscus esculentus is used as an ingredient in soups. The liber of several affords a tenacious fibre; the hemp-like substance called Sun in India is obtained from Hibiscus cannabinus. Many are beautiful objects. The hairy seeds of Gossypium furnish cotton.

TYPICAL GENERA.—Malva, Lavatera, Hibiscus.

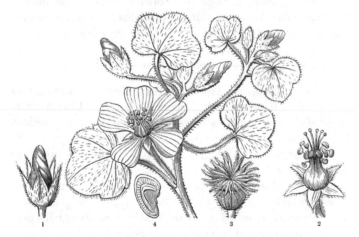

Abutilon macropodum. 1. An unexpanded flower. 2. The stamens and styles. 3. A ripe fruit, consisting of many carpels, whose upper extremities are free and radiant. 4. A section of a seed.

27.—*Tiliaceæ.* Trees or shrubs, very seldom herbaceous plants. Leaves stipulate, alternate. Flowers often small. Calyx valvate. Petals 4 or 5, usually with a little pit at their base. Stamens distinct; anthers 2-celled. Ovary single, composed of from 4 to 10 carpels; style one; stigmas as many as the carpels. Seeds several; embryo erect in the axis of fleshy albumen, with flat foliaceous cotyledons.

Uses.—Mucilaginous plants with tough fibres. The leaves of Corchorus olitorius are eaten as spinach. Corchorus capsularis furnishes a kind of coarse hemp in India. From the inner bark of Tilia Europæa Russia mats are made; its flowers, separated from the bracts, are said to be antispasmodic.

Typical Genera.—Tilia, Triumfetta, Grewia.

28.—*Dipteraceæ.* Trees. Leaves alternate, with involute vernation; stipules deciduous. Calyx 5-lobed, unequal, becoming enlarged, imbricated. Petals contorted. Stamens indefinite, distinct; anthers subulate, opening longitudinally towards the apex. Ovary without a disk, few-celled; ovules in pairs, pendulous; style single. Fruit surrounded by a calyx, having tough, leafy, enlarged, permanent divisions. Seed single, without albumen. Cotyledons crumpled.

Uses.—Tropical trees often yielding valuable timber; that called Sal, or Saul, belongs to Shorea robusta. The juice is balsamic; Dryobalanops Camphora yields Sumatra Camphor. Vateria Indica furnishes Copal. Dammer pitch is obtained from species of Shorea.

Typical Genera.—Shorea, Dipterocarpus, Vateria.

29.—*Aurantiaceæ.* Trees or shrubs, almost always smooth and filled with transparent receptacles of oil. Leaves alternate, often compound, always articulated with the petiole. Flowers usually white or green, and fragrant. Calyx urceolate or campanulate, short. Petals 3-5, inserted upon the outside of an hypogynous disk. Stamens inserted upon an hypogynous disk; filaments sometimes combined in one or several parcels. Ovary many-celled; style 1; stigma thickish. Fruit many-celled, filled with pulp. Seeds usually pendulous; raphe and chalaza distinctly marked.

Uses.—The Orange, Lemon, Lime, and Citron are species of Citrus, and are well known for the aromatic rind and pulpy flesh of their fruit. The wood is generally hard and durable. The unripe fruit of Ægle marmelos, an Indian tree, is prescribed in diarrhœa and dysentery. The leaves of the order generally are regarded as stomachic and tonic.

Typical Genera.—Citrus, Triphasia, Limonia.

30.—*Ternstromiaceæ.* Trees or shrubs. Leaves alternate,

without stipules, now and then with pellucid dots. Flowers often large and showy. Sepals 5 or 7, coriaceous, in a broken whorl, deciduous. Petals not equal in number to the sepals. Stamens numerous; monadelphous or polyadelphous. Ovary with several cells; styles filiform. Capsule 2-7-celled; usually with a central column. Seeds large, attached to the axis, very few; albumen none; cotyledons occasionally plaited.

Uses.—The Tea of Commerce consists of the leaves of Thea viridis and Bohea. Camellia oleifera yields excellent oil. The species of Camellia, common in gardens, are objects of beauty. Leaves of Kielmeyera speciosa are mucilaginous.

Typical Genera.—Camellia, Gordonia, Thea.

Kielmeyera rosea. 1. The pistil. 2. A transverse section of it. 3. A ripe fruit. 4. An embryo.

31.—*Hypericaceæ*. Herbaceous plants, shrubs or trees. Leaves opposite, entire, sometimes dotted. Flowers generally yellow. Sepals 4-5, persistent, imbricated, unequal, with glandular dots. Petals 4-5, hypogynous, twisted, oblique, often having black dots. Stamens indefinite, often polyadelphous. Styles several. Fruit a capsule or berry, of many valves and many cells. Seeds minute, indefinite; embryo straight, with no albumen.

Uses.—The juice is resinous, purgative, febrifugal or astringent in different species, according as an essential oil or a

yellow juice most abound. The latter, when concrete, resembles gamboge, of which it has the properties. Hypericum
Androsæmum and perforatum are old-fashioned vulneraries.
In Brazil a species of Hypericum is employed for a gargle
in cases of sore throat.

TYPICAL GENERA.—Hypericum, Vismia.

Hypericum floribundum. 1. An entire flower. 2. A bundle of stamens. 3. A
pistil with 3 carpels. 4. A seed laid horizontally and cut through, to show the embryo and netted testa. 5. A piece of a leaf with transparent dots.

32.—*Clusiaceæ* or *Guttiferæ*.   Trees or shrubs.   Leaves
without stipules, opposite, coriaceous.   Flowers sometimes
polygamous.   Sepals 2 to 6, persistent.   Petals hypogynous,
4 to 10.   Stamens numerous, hypogynous.   Disk fleshy, occasionally 5-lobed.   Ovary 1- or many-celled; ovules solitary,
erect, or ascending, or numerous and attached to central placentæ; style very short; stigma peltate or radiate.   Seeds
frequently nestling in pulp, often with an aril; albumen none.

USES.—Gamboge is the juice of Hebradendron cambogioides.
The delicious Malacca fruit called Mangosteen is the produce
of Garcinia mangostana.   The resinous oil Tacamahaca flows
from the root of Calophyllum Calaba.   The general properties
of the species are acrid and purgative.   They are often objects
of great beauty on account of their large flowers and handsome thick leaves.

TYPICAL GENERA.—Clusia, Garcinia.

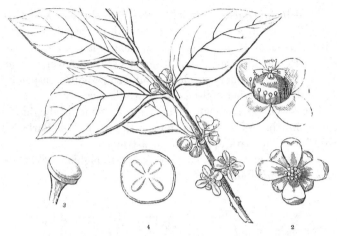

Hebradendron cambogioides. 1. A female flower, with the sterile stamens surrounding the pistil. 2. A male flower. 3. An anther, which opens by throwing off a cap, in consequence of transverse dehiscence. 4. A transverse section of the ovary.

33.—*Aceraceæ.* Trees. Leaves opposite, without stipules. Flowers small, green, often polygamous. Flowers unsymmetrical. Calyx imbricated. Petals inserted round an hypogynous disk. Stamens inserted upon an hypogynous disk, generally 8. Ovary 2-lobed; style 1. Fruit of 2 parts, which are samaroid; each 1-celled; with one or two seeds; albumen none.

Uses.—The saccharine sap of Acer saccharinum yields a kind of sugar in North America. The timber of most species is light, clean, and useful, where strength is not required.

Typical Genus.—Acer.

34.—*Æsculaceæ.* Trees or shrubs. Leaves opposite, without stipules, quinate or septenate. Racemes terminal. Flowers large, showy. Flowers unsymmetrical. Calyx campanulate, 5-lobed. Petals 4 or 5, unequal, hypogynous. Stamens 7-8, unequal. Ovary 3-celled; ovules 2 in each cell. Fruit 1- 2- or 3-valved. Seeds large, with a broad hilum; albumen none; embryo curved, germinating under ground.

Uses.—Handsome trees or bushes. Seeds filled with starch, which renders them nutritious; but it is said that they are also dangerous.

Typical Genus.—Æsculus.

35.—*Malpighiaceæ.* Small trees or shrubs, sometimes climbing. Leaves opposite, with stipules. Sepals generally with 5 pairs of large oblong conspicuous glands on the outside. Petals 5, unguiculate. Stamens seldom fewer. Ovary 1, of 3 carpels, more or less combined; ovules suspended. Fruit dry or berried. Seeds without albumen.

Uses.—Of no moment. The fruit of some Malpighias is eaten in the West Indies under the name of Barbadoes cherries. The bark appears to be astringent.

Typical Genera.—*Fruit succulent,* Malpighia. *Fruit dry and samaroid,* Banisteria.

Diplopteris paralias. 1. A flower-bud, showing the double glands of the calyx. 2. An expanded flower. 3. The carpels. 4. Ripe fruit of Ryssopteris timorensis.

36.—*Sapindaceæ.* Trees, or shrubs which often climb and have tendrils. Leaves generally compound. Flowers unsymmetrical, polygamous. Calyx imbricated. Petals hypogynous, sometimes naked, sometimes with a doubled appendage in the inside, imbricated. Disk fleshy. Stamens 8-10, rarely 5-6-7. Ovary 3-celled, the cells containing 1, 2, 3, very seldom more, ovules. Fruit sometimes capsular, sometimes samaroid, sometimes fleshy and indehiscent. Seeds usually with an aril. Albumen 0.

Uses.—Leaves and branches of some species of Magonia and Paullinia poisonous. The fruit of some Euphorias or Nepheliums, Pierardias and Hedycarya, eatable and agreeable; the former is the Longan and Litchi, which occasionally arrive in this country from China. The fruit of Sapindus saponaria and others employed instead of soap.

TYPICAL GENERA.—Sapindus, Paullinia, Serjania.

Sapindus Senegalensis. 1. An expanded flower. 2. A petal. 3. The ovaries after fertilization. 4. A vertical section of a ripe drupe, showing the embryo.

37.—*Cedrelaceæ.* Trees with timber which is usually compact, scented, and beautifully veined. Leaves alternate, without stipules. Calyx 4-5-cleft. Petals 4-5. Stamens 8-10, either united or distinct. Style and stigma simple. Seeds flat-winged.

USES.—Mahogany is the timber of Swietenia Mahagoni; the bark of that tree, of Cedrela Toona, and Soymida febrifuga, is valuable as a tonic, in careful hands; it can only be exhibited in small doses. East India Satin-wood is produced by Chloroxylon Swietenia.

TYPICAL GENERA.—Cedrela, Swietenia.

38.—*Humiriaceæ*.   Trees or shrubs.   Leaves alternate, without stipules.   Calyx 5-parted.   Stamens numerous, monadelphous; anthers with a fleshy connective extended beyond the lobes.   Ovary 5-celled; ovules 1-2, suspended; styles simple.   Fruit drupaceous.   Embryo in fleshy albumen.

USES.—The liquid yellow fragrant Balsam of Umiri flows from the wounded trunk of Humirium floribundum.   In properties it resembles Copaiva.

TYPICAL GENUS.—Humirium.

39.—*Meliaceæ*.   Trees or shrubs.   Leaves alternate, without stipules.   Sepals 3, 4, or 5.   Petals hypogynous, usually valvate.   Stamens twice as many as the petals; filaments cohering in a long tube; anthers sessile within the orifice of the tube.   Ovary with 3, 10, 12 cells; ovules suspended, 1-2 in each cell.   Fruit often 1-celled.   Seeds without albumen, not winged.

USES.—The bark of Guarea Aubletia, Trichilia cathartica, and others, purgative and emetic.   Root of Melia Azedarach anthelmintic.   Some of the tropical genera have a wholesome pleasant fruit.   Febrifugal qualities have been recognized in the Neemtree, Melia Azedarachta, and some others.

TYPICAL GENERA.—Melia, Quivisia.

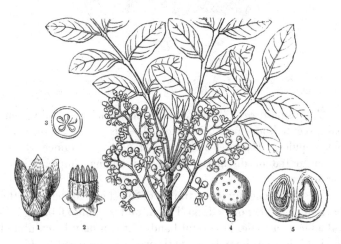

Ekebergia Senegalensis.   1. A flower.   2. The calyx and staminal tube.   3. A transverse section of the ovary.   4. A ripe fruit.   5. A vertical section of the latter.

40.—*Vitaceæ.* Scrambling, climbing shrubs, with tumid separable joints. Leaves with stipules. Flowers small, green. Calyx small, nearly entire. Petals in æstivation valvate, and often inflected at the point; stamens opposite them, inserted upon the disk. Ovary 2-celled; ovules erect, definite. Berry pulpy; albumen hard. Embryo small.

USES.—The common Vine, Vitis vinifera, is well known; besides which there are other species, in which reside similar qualities, although very inferior. The leaves of some kinds of Cissus, being acrid, are used in bringing indolent tumours to suppuration.

TYPICAL GENERA.—Vitis, Cissus, Ampelopsis.

41.—*Geraniaceæ.* Strong-scented herbs or shrubs with stipulate leaves. Stems tumid, and separable at the joints. Sepals 5, ribbed. Petals 5, unguiculate. Stamens definite, often monadelphous. Fruit of 5 elastic cocci, rolling back from a long-beaked gynobase, to which the hardened styles adhere. Seeds solitary, pendulous, without albumen. Cotyledons convolute and plaited.

USES.—The root of Geranium maculatum is a powerful astringent. Otherwise the order is of no importance, except

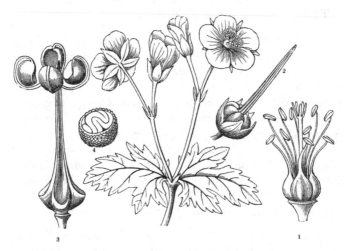

Geranium sylvaticum. 1. The stamens and style. 2. The unripe fruit surrounded by a calyx. 3. The rostrate gynobase, from which the cocci are rolling back with elasticity; one has dropped off. 4. A transverse section of a seed.

for the beautiful flowers of numerous species, especially belonging to the genus Pelargonium.

TYPICAL GENERA.—Geranium, Erodium, Pelargonium.

42.—*Balsaminaceæ.* Succulent herbaceous plants. Leaves
without stipules. Flowers usually unsymmetrical. Sepals 5,
irregular; the odd sepal spurred. Petals 4, irregular. Stamens 5. Carpels consolidated into a 5-celled ovary. Fruit
capsular, with 5 elastic valves. Seeds solitary, or numerous,
suspended; albumen none.

USES.—Unimportant. They have generally gay flowers.

TYPICAL GENERA.—Balsamina, Impatiens.

43.—*Linaceæ.* Herbaceous plants or small shrubs. Leaves
without stipules, usually alternate. Petals fugitive. Flowers
symmetrical. Sepals 3-4-5, imbricated, persistent. Petals
hypogynous, unguiculate, twisted. Stamens united in a ring.
Ovary with about as many cells as sepals. Styles equal in
number to the cells; stigmas capitate. Capsule many-celled.
Seeds in each cell single, inverted; albumen present.

USES.—The mucilaginous seeds of Linum usitatissimum
are linseed. The leaves of L. catharticum are purgative. The
tough fibre of the first is the Flax of manufacturers.

TYPICAL GENERA.—Linum, Radiola.

44.—*Oxalidaceæ.* Herbaceous plants, under-shrubs, or trees.
Leaves alternate, compound, often sensitive. Sepals 5, equal.
Petals equal, unguiculate. Stamens 10, more or less monadelphous. Styles 5; stigmas capitate. Fruit capsular, membranous, with 5 cells. Seeds few, within a fleshy integument,
which expels the seeds with elasticity. Embryo long, taper.
Albumen between cartilaginous and fleshy.

USES.—They are generally acid in a high degree. The
Blimbing and Carambola, acid fruits of the Indian Archipe-
lago, are the produce of the genus Averrhoa. The roots of
Oxalis Deppei form an agreeable esculent.

TYPICAL GENERA.—Oxalis, Averrhoa.

45.—*Pittosporaceæ.* Leaves simple, alternate, without stipules. Shrubs, trees, or half herbaceous plants, sometimes
twining. Sepals deciduous, imbricated. Petals hypogynous,

imbricated. Stamens 5. Ovary single, many-seeded. Fruit capsular or berried, with many-seeded cells which are sometimes incomplete. Albumen fleshy.

Uses.—Unimportant. The species are resinous.

Typical Genera.—Pittosporum, Sollya, Billardiera.

46.—*Rutaceæ.* Trees or shrubs (or herbs). Leaves exstipulate, dotted. Flowers often very gay. Flowers hermaphrodite, sometimes irregular. Sepals 4-5. Petals sometimes combined. Stamens definite, on the outside of a cup-like disk. Ovary few-celled; ovules 2-4; style single, occasionally divided near the base, always separable into its component parts as the fruit approaches maturity. Fruit capsular, separating into carpels when ripe. Embryo with or without albumen; radicle superior.

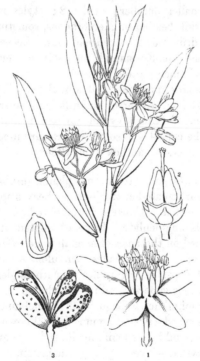

Eriostemon myoporoides. 1. A complete flower. 2. The ovary, seated in a cup-shaped disk, surrounded by a calyx. 3. The ripe fruit, separated spontaneously into its component carpels. 4. A vertical section of a seed, showing the embryo lying in the midst of albumen.

Uses.—The powerfully scented oil possesses active properties. Ruta graveolens, common Rue, is anthelmintic, sudorific, and emmenagogue. Various species of Barosma, called at the Cape of Good Hope Bucku, are powerful antispasmodics. The bark of Cusparia febrifuga, called Angostura bark, is a valuable febrifuge ; and that of many other American trees of the order seems to possess the same quality.

Typical Genera.—Ruta, Boronia, Dictamnus,—Correa is remarkable for having a monopetalous corolla.

47.—*Xanthoxylaceæ.* Trees or shrubs. Leaves without stipules, with pellucid dots. Flowers unisexual. Calyx in 3, 4, or 5 divisions. Petals usually longer than the calyx, convolute. Stamens equal to the petals in number, or twice as many. Ovary of the same number of carpels as there are petals, or a smaller number ; ovules 2 ; styles more or less combined. Fruit berried or membranous, sometimes consisting of several drupes or 2-valved capsules. Seeds solitary or twin, pendulous, usually smooth and shining ; embryo lying within fleshy albumen ; radicle superior.

Uses.—Aromatic, pungent, and stimulant. Xanthoxylum Clava Herculis is a powerful sudorific and aperient. The bark of Brucea, of Xanthoxylum caribæum, and others, is febrifugal. The capsules of some Fagaras are used as pepper.

Typical Genera.—Xanthoxylum, Ptelea.

48.—*Zygophyllaceæ.* Herbaceous plants, shrubs, or trees ; branches often articulated at the joints. Leaves opposite, with stipules, not dotted. Flowers hermaphrodite. Calyx convolute. Petals unguiculate. Stamens dilated at the base, sometimes placed on the back of a small scale. Ovary with a disk, and 4 or 5 cells ; ovules pendulous or erect ; style simple. Fruit capsular, rarely fleshy, with angles or wings. Seeds few ; radicle superior ; albumen whitish.

Uses.—Zygophyllum Fabago is an anthelmintic. Guaiacum yields the wood called Lignum Vitæ, known in turnery for its hardness, and in medicine for its sudorific qualities.

Typical Genera.—Zygophyllum, Guaiacum.

49.—*Simarubaceæ.* Trees or shrubs. Leaves without

stipules, alternate, without dots. Flowers hermaphrodite, or unisexual. Calyx in 4 or 5 divisions. Petals longer; æstivation twisted. Stamens arising from the back of an hypogynous scale. Ovary 4- or 5-lobed, upon a stalk, each cell with 1 suspended ovule; style simple. Fruit indehiscent; embryo without albumen.

Uses.—The wood intensely bitter. The root of Simaruba amara, used as a tonic, is bitter, purgative, and emetic. The wood of Picræna excelsa furnishes the Quassia chips of the shops.

Typical Genera.—Quassia, Simaruba.

50.—*Coriariaceæ.* Shrubs with square branches. Leaves opposite, simple, entire, ribbed. Flowers herbaceous, hermaphrodite, monœcious or diœcious. Calyx 5-parted. Petals 5, fleshy, with an elevated keel. Stamens 10. Ovary 5-celled, 5-angled; style none; stigmas 5, subulate. Ovules solitary. Carpels 5, drupaceous, indehiscent, 1-seeded, sometimes surrounded by the enlarged petals. Albumen none.

Uses.—The fruit of Coriaria myrtifolia is poisonous; the leaves are used for dyeing black, and for adulterating Alexandrian Senna.

Typical Genus.—Coriaria.

SUBCLASS II. CALYCIFLORÆ.

51.—*Celastraceæ.* Shrubs or trees. Leaves simple. Flowers in axillary cymes, minute. Sepals 4 or 5, imbricated, inserted into the margin of an expanded torus. Petals imbricate. Stamens alternate with the petals, inserted into the disk. Disk large, expanded, flat, closely surrounding the ovary. Ovary with 3 or 4 cells; ovules ascending; fruit capsular or drupaceous; seeds often with an aril; albumen fleshy.

Uses.—Sub-acrid, but apparently unimportant plants in a medicinal point of view. A yellow die is obtained from the bark of Euonymus tingens in India.

Typical Genera.—Celastrus, Euonymus.

52.—*Staphyleaceæ.* Shrubs. Leaves opposite, pinnate, with both common and partial stipules. Sepals 5, coloured, imbricated. Petals 5, imbricated. Stamens alternate with the petals, perigynous. Disk large, urceolate. Ovary 2- or

3-celled, superior; ovules erect; styles 2 or 3, cohering. Fruit membranous or fleshy. Seeds with a bony testa and no aril; hilum large; albumen none.

Uses.—Staphylea pinnata and trifolia are cultivated as ornamental shrubs under the name of Bladder-nuts, because their nut-like seeds are enclosed in a bladdery seed-vessel.

Typical Genus.—Staphylea.

53.—*Rhamnaceæ*. Trees or shrubs. Leaves alternate, with minute stipules. Flowers axillary or terminal, minute. Calyx 4-5-cleft, valvate. Petals distinct, inserted into the orifice of the calyx. Stamens definite, opposite the petals, to which they are equal in number. Ovary superior, or half-superior, 2-3- or 4-celled; ovules solitary, erect; fruit a capsule, or more frequently a berry; albumen fleshy, in very small quantity; embryo with large flat cotyledons, and a short inferior radicle.

Uses.—The berries of Rhamnus Frangula, catharticus, and others, are active purgatives. When ripe, those of some species, especially R. catharticus and infectorius, yield a yellow dye. The fruit of Zizyphus communis is the Jujube of the shops, and that of the Z. Lotus gave their name to the Lotophagous nation of antiquity; all the fruit of that genus seems harmless; Z. Chinensis, indeed, is cultivated in China as the apple is with us. The bark of Ceanothus americanus and some others is astringent, and has been employed in diarrhœa.

Typical Genera.—Rhamnus, Paliurus, Ceanothus.

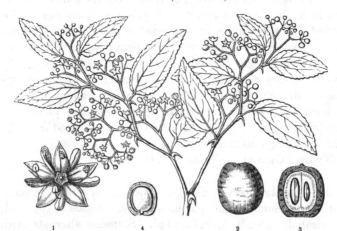

Zizyphus Baclei. 1. A flower seen from above. 2. A fruit. 3. The same cut vertically. 4. A seed divided vertically.

54.—*Anacardiaceæ.* Trees or shrubs, with a resinous caustic juice, becoming black in drying. Leaves alternate, without pellucid dots. Flowers small, green, unisexual. Calyx small. Petals perigynous, imbricated. Stamens usually definite. Disk fleshy, hypogynous. Carpel simple; styles 1 or 3, occasionally 4; ovule solitary, attached by a cord to the bottom of the cell. Fruit indehiscent. Seed without albumen.

USES.—A hard, black, acrid varnish is obtained from Semecarpus Anacardium and Melanorhœa usitatissima. The Cashew nut, whose eatable kernel is surrounded by a rind full of resinous acrid oil, is produced by Anacardium occidentale. Rhus toxicodendron and radicans are dangerous acrid poisons. Rhus Coriaria, Cotinus, and some others, are astringents. Rhus Typhinum, and some others, are cultivated as plants of ornament. Pistacia Atlantica and Lentiscus yield mastich; and P. Terebinthus, Scio turpentine.

TYPICAL GENERA.—Rhus, Pistacia.

Pistacia Atlantica. 1. Female flowers. 2. An ovary. 3. The same cut open to show the ovule. 4. A ripe fruit opened to show the seed.

55.—*Fabaceæ* (or *Leguminosæ*). Herbaceous plants, shrubs, or trees. Leaves alternate; petiole tumid at the base. Stipules 2. Flowers usually showy. Calyx inferior, the segments often unequal, and variously combined. Petals either papilionaceous or regularly spreading. Stamens definite or indefinite, perigynous, or hypogynous. Ovary simple, superior. Fruit a legume. Seeds destitute of albumen.

A very large natural order, of which there are 3 principal divisions:—

Division 1.—*Papilionaceæ.*   Flowers papilionaceous (340).

USES.—The Locust-tree, Laburnum, and Sissoo, a species of Dalbergia, yield valuable wood.   The roots of Glycyrhiza glabra are liquorice.   Peas, Beans, Kidney Beans, Vetches, and other sorts of pulse, are articles of food.   Clover, Suckling, Melilot, Lucerne, Medick, Saintfoin, and others, are fodder plants.   Indigo is furnished by various plants, especially Indigofera tinctoria.   A kind of manna oozes from Alhagi Maurorum.   Cowhage consists of the stinging hairs on the pods of Mucuna pruriens.   Certain Astragali yield gum Tragacanth.   The seeds of Laburnum and several others are narcotic; as also is the root of Piscidia Erythrina, the tincture of which is said to be more powerful than laudanum.

TYPICAL GENERA.— Cytisus, Lathyrus, Colutea.

Adenocarpus frankenioides.   1. The standard, wings, and keel split open.   2. The stamens.   3. A cross section of a seed.   4. A legume, with a portion of one of the valves turned back.

Division 2.—*Cæsalpinieæ.*   Petals regularly spreading, imbricated.   Stamens perigynous.

USES.—Senna is the foliage of different species of Cassia. The Tamarind fruit comes from Tamarindus Indica.   The pods of the Carob-tree (Ceratonia Siliqua) are highly nutritious.   Hæmatoxylon Campeachianum yields logwood; Cæsalpinia Brasiliensis, Brazil wood.

TYPICAL GENERA.—Cassia, Bauhinia.

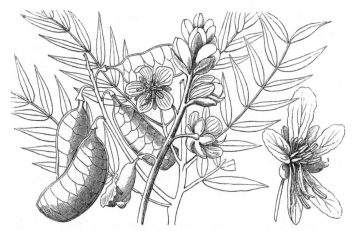

Cassia acutifolia.   1. A flower somewhat magnified.

Division 3.—*Mimoseæ*.   Sepals and petals valvate.   Stamens hypogynous.

Uses.—Gum Arabic oozes from various species of Acacia, especially A. Verek and arabica.   Catechu is obtained by boiling the bark of A. Catechu; and many kinds are employed for tanning purposes.   Several are fine timber trees.   Finally, the leaves and branches of some kinds are poisonous.   The spongy stems of Desmanthus natans supply a coarse kind of rice paper.   Most are objects of great beauty.

Typical Genera.—Acacia, Mimosa, Inga.

Acacia Verek.   1. A flower magnified.   2. The pistil.   3. A section of the same.   4. Half a seed.

56.—*Rosaceæ*. Trees, shrubs, or herbaceous plants. Leaves alternate, usually with conspicuous stipules, more frequently compound than simple. Flowers large, showy, arranged variously, but in most cases terminal. Calyx lined with a disk. Petals equal. Stamens usually indefinite. Carpels solitary or several, disunited or consolidated. Styles distinct, and more or less obliquely placed upon the ovary. Fruit various. Seeds without albumen. Embryo straight.

Division 1.—*Roseæ*. Tube of calyx fleshy, and covering over the achænia with a false pericarp.

Uses.—Fruit astringent. Petals fragrant and astringent. Flowers in all cases beautiful.

Typical Genus.—Rosa.

Division 2.—*Potentilleæ*. Carpels numerous, superior, indehiscent.

Uses.—Usually gay flowers. The fruit of Fragaria is the Strawberry, of Rubus the Bramble and the Raspberry. The roots of Tormentils and some Geums and Potentillas are astringent, and have been used as febrifuges.

Typical Genera.—Rubus, Fragaria.

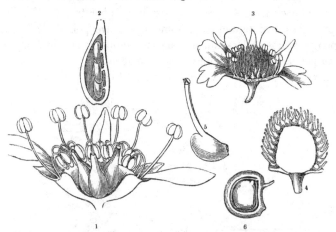

1. Spiræa Aruncus, flower cut open. 2. A section of an ovary. 3. Part of flower of Fragaria Indica. 4. A vertical section of the half-ripe receptacle, covered with carpels. 5. A single carpel. 6. A section of a ripe carpel, with the seed inside.

Division 3.—*Spiræeæ*. Carpels few, 2-valved.

Uses.—Roots of Gillenia emetic, of Spiræa ulmaria tonic.

Typical Genus.—Spiræa.

Division 4.—*Amygdaleæ.* Carpel single, a drupe.

Uses.—The fruit of the Peach, Nectarine, Almond, Plum, Cherry, and Apricot, are produced by various species. Many are of great beauty on account of their gay flowers. Hydrocyanic acid is yielded by the leaves of all, especially of the Prunus Laurocerasus, or common Laurel. The bark of Prunus Coccomilia and some others is febrifugal.

Typical Genera.—Prunus, Amygdalus.

Division 5.—*Pomeæ.* Carpels adhering to the calyx.

Uses.—Beautiful trees or bushes, bearing a fruit which is, in the majority of species, eatable. Apples, Pears, Quinces, Medlars, Services, are the produce of different species. The wood is usually very hard. The Hawthorn is a valuable material for fences.

Typical Genera.—Pyrus, Cratægus.

Division 6.—*Sanguisorbeæ.* Flowers often unisexual. Petals none. Tube of the calyx hardened.

Uses.—Astringents of little importance. Common Burnet used for sheep pasture is Sanguisorba officinalis.

Typical Genera.—Alchemilla, Sanguisorba, Poterium.

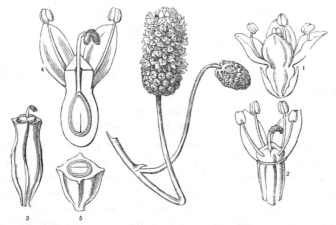

Sanguisorba officinalis. 1. A flower with a pair of bracts. 2. The same with half the calyx cut away. 3. A ripe fruit, from which the calyx has been removed. 4. A vertical section of fruit and calyx. 5. Transverse section of a fruit.

57.—*Amyridaceæ.* Resinous trees or shrubs. Leaves compound, with pellucid dots. Flowers axillary and terminal, panicled. Calyx minute. Petals 4-5, imbricated. Stamens definite. Ovary superior, 1-celled; stigma capitate; ovules pendulous. Fruit indehiscent, glandular. Seed without albumen; radicle superior.

USES.—Fragrant resinous shrubs. Myrospermum toluiferum yields Balsam of Tolu; and Gum Elemi comes from some species of Amyris. Balsam of Copaiva is furnished by different species of Copaifera.

TYPICAL GENERA.—Amyris, Copaifera.

58.—*Chrysobalanaceæ.* Trees or shrubs. Leaves simple, alternate, stipulate, with veins that run parallel with each other from the midrib to the margin. Calyx 5-lobed. Petals more or less irregular, either 5 or none. Stamens definite or indefinite, usually irregular. Ovary superior, solitary, cohering more or less on one side with the calyx; ovules erect. Fruit, a drupe with 1 or 2 cells. Seed solitary, erect. Embryo with no albumen.

USES.—The fruit of Chrysobalanus Icaco is the Cocoa Plum of the West Indies. The general properties appear to be innocuous.

TYPICAL GENERA.—Hirtella, Chrysobalanus.

59.—*Calycanthaceæ.* Shrubs with square stems. Leaves opposite, simple, without stipules. Sepals and petals confounded, indefinite, combined in a fleshy tube. Stamens inserted in a fleshy rim. Anthers adnate, turned outwards. Ovaries several, simple; ovules solitary. Nuts enclosed in the fleshy tube of the calyx, 1-seeded. Albumen none; cotyledons convolute.

USES.—The flowers are fragrant, but of no known use.

TYPICAL GENERA.—Calycanthus, Chimonanthus.

60.—*Lythraceæ.* Herbs, rarely shrubs. Branches frequently 4-cornered. Leaves without stipules. Flowers in many cases showy. Calyx tubular. Petals inserted between the lobes of the calyx, very deciduous. Stamens inserted into the tube of the calyx below the petals. Ovary 2- or 4-celled; style filiform; capsule membranous, covered by the calyx, usually 1-celled. Seeds numerous, without albumen.

Uses.—The Lagerströmias and Lythrums, as well as some others, are species of great beauty. Lythrum Salicaria has been employed in diarrhœa on account of its astringency. Lawsonia inermis furnishes the Henné with which oriental women stain their nails. A few are acrid.

Typical Genera.—Lythrum, Ammannia.

61.—*Combretaceæ.* Trees or shrubs. Leaves without stipules. Flowers generally showy. Calyx 4- or 5-lobed, deciduous. Stamens twice as many as the segments of the calyx, or three times as many. Ovary 1-celled, with from 2 to 4 ovules, hanging from the apex of the cavity. Seed without albumen; cotyledons usually convolute.

Uses.—Many species are astringent, and are used by the tanners in the countries where they occur. Myrobalan nuts, also tonics, are produced by Terminalia bellerica. The kernels of T. Catappa are eaten like almonds.

Typical Genera.—Combretum, Conocarpus.

62.—*Melastomaceæ.* Trees, shrubs, or herbaceous plants. Leaves opposite, with several ribs. Flowers generally purple and very handsome. Calyx cohering with the angles of the ovary. Petals twisted in æstivation; filaments curved downwards in æstivation; anthers 2-celled, elongated beyond the insertion of the filament (see fig. 211, p. 46). Ovary with several cells, and indefinite ovules. Pericarp with placentæ attached to a central column. Seeds innumerable.

Uses.—Of no importance. The species appear harmless; their fruit, when succulent, is eatable.

Typical Genera.—Rhexia, Melastoma, Lasiandra.

63.—*Philadelphaceæ.* Shrubs. Leaves deciduous, opposite, without dots or stipules. Flowers usually white. Calyx persistent, having from 4 to 10 divisions. Petals convolute, imbricate. Stamens definite. Styles indistinct, or consolidated; stigmas several. Capsule with 4 to 10 cells, many-seeded. Seeds scobiform; aril loose, membranous. Albumen fleshy.

Uses.—Merely known as plants of ornament, and sometimes of fragrance. The rough leaves of Deutzia are said to be used by the Japanese as a polishing material.

Typical Genera.—Philadelphus, Deutzia.

64.—*Myrtaceæ.* Trees or shrubs. Leaves with transparent dots, and often with a vein running parallel with their
margin. Calyx 4- or 5-cleft, sometimes like a cap. Petals
quincuncial or wanting. Stamens indefinite; anthers ovate,
small. Ovary 1-2-4-5-6-celled. Fruit either dry or fleshy.
Seeds definite or indefinite; embryo without albumen.

The principal divisions are the following:—

Division 1.—*Myrteæ.* Fruit 2- or more celled, fleshy.

Division 2.—*Leptospermeæ.* Fruit 2- or more celled, capsular.

Division 3.—*Chamælauciæ.* Fruit 1-celled.

Uses.—The spices called Cloves and Pimento are the dried
flowers of Caryophyllus aromaticus and the dried fruit of
Eugenia Pimenta. The New Holland Eucalypti contain a
great quantity of tannin in their bark. Cajeputi oil is obtained
from Melaleuca Cajeputi. The bark of the Pomegranate
root is an anthelmintic. Almost all the species are beautiful
either in foliage or flower.

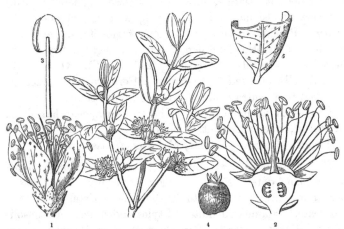

Eugenia tuberculata. 1. A flower. 2. The same divided vertically. 3. A stamen. 4. A ripe fruit. 5. A leaf with the dots upon it.

65.—*Onagraceæ.* Herbaceous plants or shrubs. Leaves
alternate or opposite. Flowers generally showy. Calyx tubular, 4-lobed, valvate. Petals regular, with a twisted æstivation. Stamens 2, 4, or 8, inserted into the calyx. Styles

consolidated. Stigma 4-lobed. Fruit many-seeded, with four cells. Seeds without albumen.

Division 1.—*Œnothereæ*. Petals 4. Fruit a capsule. Seeds numerous.

USES.—These are gay flowered plants, of no known use.

TYPICAL GENERA.—Œnothera, Epilobium.

Ludwigia Jussiæoides. 1. A flower with two sepals and all the petals cut off. 2. A calyx and inferior ovary. 3. A transverse section of the ovary. 4. A seed with the distinct raphe. 5. An embryo extracted.

Division 2.—*Fuchsieæ*. Petals 4. Fruit a succulent berry.

USES.—Unknown. Beautiful bushes.

TYPICAL GENUS.—Fuchsia.

Division 3.—*Circæeæ*. Petals 2, 4, or none. Stamens 1 or 2. Fruit a capsule.

USES.—Unknown.

TYPICAL GENERA.—Circæa, Lopezia.

66.—*Cercodiaceæ*. Small shrubs, or obscure herbaceous plants. Calyx minute. Petals small or none. Stamens 1-8. Ovary 1-4-celled. Styles distinct. Ovules pendulous, anatropous (460). Fruit nut-like, usually crowned by the rim of the calyx. Seed pendulous, with a small quantity of fleshy albumen.

Division 1.—*Hippurideæ*. Calyx obsolete. Petals none. Stamen 1.

USES.—Unknown. Obscure weeds.

TYPICAL GENUS.—Hippuris.

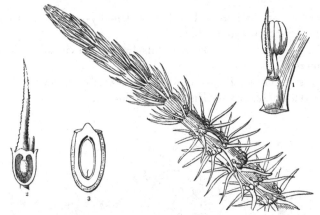

Hippuris vulgaris. 1. A complete flower. 2. A section of the pistil, showing the position of the ovule. 3. A section of the ripe fruit and seed.

Division 2.—*Halorageæ*. Calyx toothed. Petals present. Stamens more than one.

Loudonia aurea. 1. A complete flower. 2. A section of the ovary.

Uses.—Unknown.
Typical Genera.—Myriophyllum, Loudonia.

67.—*Loasaceæ.* Herbaceous plants, hispid, with pungent hairs. Leaves without stipules. Flowers generally showy, white or yellow. Calyx 5-parted. Petals 5 or 10, hooded, with an inflexed æstivation; the interior often much smaller. Stamens indefinite. Ovary with several parietal placentæ, or with a free central lobed one. Fruit capsular or succulent. Seeds numerous, without aril; embryo in axis of fleshy albumen.
Uses.—Unknown. Usually handsome plants.
Typical Genera.—Loasa, Bartonia.

68.—*Cucurbitaceæ.* Annual or perennial herbs. Stem climbing by tendrils. Leaves palmated, or with palmate ribs, covered with asperities. Flowers white, yellow, or brownish red, unisexual. Calyx 5-toothed. Corolla 5-parted, scarcely distinguishable from the calyx, with strongly reticulated veins. Stamens 5, either distinct, or cohering in three parcels; anthers sinuous. Ovary with 3 parietal placentæ; stigmas very thick, velvety or fringed. Fruit more or less succulent. Seeds flat, in an aril; embryo flat, with no albumen.
Uses.—The Gourd, Melon, Cucumber, Pumpkin, Vegetable Marrow, and Squash, are the fruits of various species, in all which an acrid purgative principle is diffused; which, when concentrated, as in the Bottle Gourd, the Colocynth, and the Bryony, becomes dangerous, unless administered with skill, when it is a useful medicine. Elaterine, or Elatine, an extremely poisonous principle, is found in the Spirting Cucumber, Momordica Elaterium. The seeds are nutty and harmless.
Typical Genera.—Cucumis, Bryonia, Momordica.

69.—*Passifloraceæ.* Usually climbing by means of tendrils. Leaves alternate, with leafy stipules. Flowers often enclosed in an involucre. Sepals 5, their tube lined with filamentous processes. Petals 5. Stamens monadelphous. Ovary stalked, 1-celled; styles 3; stigmas simple, clavate. Fruit with 3 polyspermous placentæ. Seeds with a brittle sculptured testa. Embryo in fleshy albumen.
Uses.—The fruit of Passiflora quadrangularis, the Granadilla, of P. edulis, and several others, contains a pleasant sub-

K

acid pulp, on account of which they are served up at dessert.
The root of the first species is emetic and narcotic; and similar properties are ascribed to that of P. rubra, which is called
in Jamaica Dutchman's laudanum.   P. fœtida has some
reputation as an emmenagogue.
TYPICAL GENERA.—Passiflora, Tacsonia.

70.—*Turneraceæ.* Herbaceous plants. Leaves alternate,
without stipules, with occasionally two glands at the apex
of the petiole. Calyx often coloured, with 5 lobes, imbricated. Petals 5, equal, twisted. Stamens distinct. Ovary
with 3 placentæ; ovules indefinite; styles 3 or 6, cohering
more or less. Capsule 3-valved, the valves bearing the placentæ in the middle. Seeds with a thin aril on one side;
embryo in the middle of fleshy albumen.
USES.—Unknown.
TYPICAL GENUS.—Turnera.

71.—*Portulacaceæ.* Succulent shrubs or herbs. Leaves
without stipules, or sometimes with membranous ones. Flowers usually ephemeral. Sepals 2. Petals generally 5. Stamens inserted irregularly into the calyx, or hypogynous,
variable in number. Ovary 1-celled; style single, stigmas
several. Capsule 1-celled. Seeds attached to a central placenta. Embryo curved round the albumen.
USES.—Insipid plants, occasionally employed as esculents,
as in the case of Portulaca oleracea, the common Purslane.
TYPICAL GENERA.—Calandrinia, Montia.

72.—*Illecebraceæ.* Herbaceous or half shrubby plants, with
scarious stipules. Flowers minute, with scarious bracts. Sepals 3, 4, or 5. Petals minute. Stamens definite. Ovary
superior; styles 2-5. Fruit dry, 1-3-celled. Seeds upon a
central placenta; embryo on one side of the albumen.
USES.—Unimportant weeds; said to be slightly astringent.
TYPICAL GENERA.—Herniaria, Illecebrum.

73.—*Scleranthaceæ.* Small herbs. Leaves opposite, without stipules. Flowers axillary, sessile, minute, hermaphrodite. Calyx 4- or 5-toothed. Stamens from 1 to 10. Ovary
simple, superior, 1-seeded. Fruit a utricle enclosed within the

hardened calyx. Seed pendulous from a funiculus; embryo cylindrical, curved round farinaceous albumen.

Uses.—Unknown. Mere weeds.

Typical Genus.—Scleranthus.

74.—*Crassulaceæ.* Succulent herbs or shrubs. Stipules none. Flowers usually in cymes, showy. Sepals from 3 to 20. Petals either distinct or cohering. Stamens inserted with the petals. Hypogynous scales usually several, 1 at the base of each carpel. Ovaries of the same number as the petals, opposite to which they are placed. Fruit of several follicles, opening on their face. Seeds variable in number.

Uses.—Sempervivum tectorum, and many others, are refrigerants and somewhat acrid. Some are plants of considerable beauty, and capable of growing in the most exposed and sunburnt places. Sempervivum glutinosum is used to impregnate the water in which the fishermen of Madeira steep their nets, in order to render them durable.

Typical Genera.—Sempervivum, Sedum.

75. — *Mesembryaceæ* or *Ficoideæ.* Succulent shrubs or herbs. Flowers showy, opening only under bright sunshine. Sepals definite, succulent. Petals indefinite, linear. Stamens indefinite. Ovary many-celled. Stigmas numerous. Capsule many-celled, with a starry dehiscence. Embryo curved or spiral, on the outside of mealy albumen.

Uses.—Mesembryanthemum emarcidum, the Hottentot's fig, when bruised and fermented, becomes narcotic, and is used like tobacco. M. crystallinum and nodiflorum are collected in the countries where they grow wild, for the sake of the alkali they contain.

Typical Genus.—Mesembryanthemum.

76.—*Cactaceæ.* Succulent shrubs, usually destitute of leaves, and with spinous buds. Flowers usually very handsome. Sepals indefinite, confounded with the petals. Stamens indefinite; filaments long, filiform. Ovary inferior, 1-celled, with numerous parietal placentæ; stigmas numerous. Fruit succulent. Seeds without albumen.

Uses.—The fruit is eaten under the name of Indian figs.

Typical Genera.—Cereus, Mammillaria.

77.— *Grossulaceæ.* Bushes with alternate leaves, membranous stipules, and a plaited vernation, often spiny. Flowers in axillary racemes. Calyx superior, 4- or 5-parted, regular. Petals 5, minute. Stamens 5. Ovary inferior, 1-celled, with 2 parietal placentæ. Berry 1-celled, many-seeded; embryo minute, in horny albumen.

Uses.—Ribes rubrum is the common garden Currant, R. nigrum the Black Currant, and R. Grossularia the Gooseberry, all well-known fruits. Many have beautiful flowers.

Typical Genus.—Ribes.

78.—*Saxifragaceæ.* Herbaceous plants. Leaves simple, with or without stipules. Calyx superior or inferior. Petals 5, or none. Stamens 5-10, perigynous or hypogynous; anthers bursting longitudinally. Disk hypogynous or perigynous, rarely consisting of 5 scales. Ovary 1-celled, with two parietal placentæ. Styles 2, formed from extended points of the ovary. Fruit membranous, with two divaricating lobes. Seeds numerous, very minute. Embryo taper, in the axis of fleshy albumen.

Uses.—Heuchera Americana, and some others, have astringent roots. Many are pretty flowers.

Typical Genera.—Saxifraga, Heuchera.

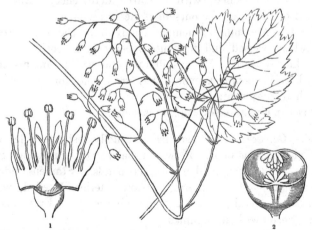

Heuchera glabra. 1. A flower split open, showing the two styles. 2. A transverse section of an ovary.

79. — *Escalloniaceæ.* Shrubs with alternate, toothed, glandular, exstipulate leaves. Flowers showy. Calyx 5-toothed. Petals forming a tube, but finally separating ; æstivation imbricated. Stamens definite. Disk conical, epigynous. Ovary 2-celled, with two large polyspermous placentæ in the axis ; style simple ; stigma 2-lobed. Fruit capsular, splitting by the separation of the cells at their base. Seeds minute ; embryo in oily albumen.

USES.—Unknown.

TYPICAL GENUS.—Escallonia.

80.—*Hamamelaceæ.* Shrubs. Leaves alternate, with deciduous stipules. Flowers sometimes unisexual. Calyx in 4 pieces. Petals 4, linear. Stamens 8 ; 4 being sterile. Ovary 2-celled ; styles 2 ; ovules solitary, pendulous. Fruit capsular. Embryo in the middle of horny albumen.

USES.—Unknown.

TYPICAL GENERA.—Hamamelis, Fothergilla.

81.—*Araliaceæ.* Trees, shrubs, or herbaceous plants, with the habit of Apiaceæ. Calyx entire or toothed. Petals 5-10. Stamens equal to the petals or twice as many, arising from without an epigynous disk. Ovary with more cells than two. Fruit succulent or dry, consisting of several 1-seeded cells. Seeds pendulous. Embryo minute, in copious albumen.

USES.—Panax quinquifolium forms the root Ginseng, regarded by the Chinese as a powerful stimulant. A sort of Sarsaparilla is prepared in North America from Aralia nudicaulis. Common Ivy, Hedera Helix, has irritating leaves.

TYPICAL GENERA.—Hedera, Aralia.

82.—*Cornaceæ.* Trees or shrubs, seldom herbs. Leaves (except in one species) opposite, entire or toothed. Flowers occasionally diœcious. Sepals 4. Petals 4, oblong, broad, valvate. Stamens 4, alternate with the petals. Drupe crowned by the calyx, 2-celled. Seeds pendulous, solitary. Albumen fleshy.

USES.—Cornus mascula, the Cornelian Cherry, and some others, produce a succulent eatable fruit of bad quality. C. florida and sericea have a powerfully tonic bark.

TYPICAL GENERA.—Cornus, Aucuba.

83. — *Apiaceæ* or *Umbelliferæ*. Herbaceous plants with fistular stems. Flowers in umbels. Calyx entire or 5-toothed. Petals 5, usually inflexed at the point. Stamens 5, alternate with the petals. Ovary 2-celled. Styles 2, diverging; disk double, epigynous. Fruit consisting of 2 carpels, or mericarps, separable from a common axis. Seed solitary, pendulous. Embryo minute, at the base of horny albumen.

Uses.—The Carrot, Parsnip, Parsley, Fennel, Skirret, and others, are eatable. Celery is poisonous when wild, bland if cultivated. Many species are dangerous poisons, as Œnanthe crocata, Cicuta virosa, Conium maculatum, Æthusa Cynapium; others have aromatic carminative fruits, as Caraway, Dill, Coriander, Anise. Assafœtida, Ammoniacum, Opopanax, fœtid gum resins, exude from certain Oriental species.

Typical Genera. — Pastinaca, Carum, Petroselinum, Daucus.

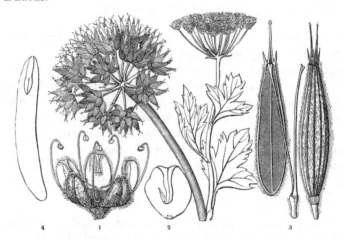

Athamanta cervariæfolia. 1. A separate flower, with hairy petals. 2. A petal by itself. 3. A ripe fruit with the two carpels or mericarps separating from the double carpopod or axis. 4. A seed deprived of its integuments, and divided vertically, so as to show the position of the embryo.

The genera of this large and difficult order being characterized very much by peculiarities in their fruit, the following cut is intended to explain the principal terms employed in speaking of them.

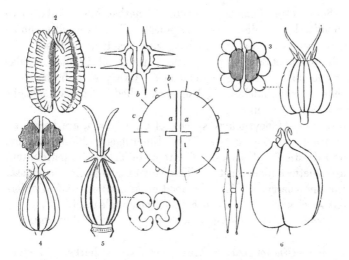

1. Is an ideal plan of a fruit divided transversely ; *a a* is the commissure, or plane of contact of the mericarps ; *b b* primary ridges ; *c c* secondary ridges. 2. Is a view of the back and section of the fruit of Laserpitium Siler ; each mericarp has the secondary ridges winged, the primary obsolete ; there are two vittæ on the commissure, and one under each secondary ridge ; these vittæ, which are cavities containing oil, are represented by dots ; the albumen is solid. 3. Sclerosciadium humile ; the primary ridges are corky ; there are no secondary ridges ; the vittæ alternate with the primary ridges, and there is one at each edge of the commissure ; the albumen is solid. 4. Discopleura capillacea ; there are 5 very small primary juga, the two lateral of which are in contact with a thickened accessory margin ; there are 2 vittæ on each face of the commissure, and one between each primary ridge ; the albumen is solid. 5. Echinophora spinosa ; albumen involute ; vittæ alternate with the primary ridges. 6. Compressed fruit of Diposis saniculæfolia ; the commissure is very narrow ; there are 5 minute primary ridges ; one along the back, one along each edge, and two on the inflexed side ; the albumen is solid.

SUBCLASS III. COROLLIFLORÆ.

84. — *Loranthaceæ.*   Parasitical half-shrubby plants. Leaves opposite, without stipules.   Flowers either very long and tubular, or small and green.   Calyx with 2 bracts at the base.   Corolla with 3, 4 or 8 petals, more or less united at the base, valvate ; stamens opposite to them.   Ovary 1-celled ; ovule erect.   Fruit succulent.   Seed solitary ; embryo cylindrical, longer than the fleshy albumen.

USES.—Bark astringent ; that of Loranthus tetrandrus is employed in Chili for a black dye.   Miseltoe is Viscum album.

TYPICAL GENERA.—Viscum, Loranthus.

85. — *Caprifoliaceæ*. Shrubs or herbaceous plants, with opposite leaves, destitute of stipules. Flowers usually showy and fragrant. Calyx 4-5-cleft, with bracts at its base. Corolla monopetalous or polypetalous, rotate or tubular, regular or irregular. Stamens epipetalous. Ovary with from 1 to 5 cells. Fruit indehiscent, 1 or more celled. Embryo straight in fleshy albumen.

Uses.—Honeysuckles, species of Caprifolium, are beautiful, fragrant, twining shrubs. The Elder has sudorific flowers, and drastic fœtid leaves. The roots of Triosteum perfoliatum are emetic and cathartic. The fruit of Symphoria racemosa, the Snowberry, is a favourite food of pheasants; that of different species of Viburnum is eatable, but unpleasant.

Typical Genera.—Sambucus, Caprifolium, Viburnum.

86. — *Cinchonaceæ*. Trees, shrubs, or herbs. Leaves simple, opposite or verticillate, with interpetiolary stipules, which are simple, bifid, or multifid, and form one of the principal characteristics of the order. Inflorescence extremely varied. Calyx simple. Corolla tubular, regular, valvate, or imbricated. Stamens all on the same line, alternate with the lobes of the corolla. Ovary surmounted by a disk; ovules numerous or few. Fruit either splitting, or indehiscent and dry, or succulent. Seeds definite or indefinite; embryo small, surrounded by horny albumen.

Uses.—Foremost among the useful species of this large order stand the species of Cinchona, whose bark is so valuable on account of its tonic febrifugal qualities; in this respect a large number of other genera correspond, among which are Buena, Remija, Portlandia, and Exostema. Others are powerful emetics; as Cephaelis Ipecacuanha, whose roots form the best Ipecacuanha of the shops; Richardsonia scabra, and several species of Manettia, Chiococca, and Spermacoce. A few have the emetic principle so concentrated as to be dangerous poisons, as Randia dumetorum. Coffee is the horny albumen of Coffea Arabica.

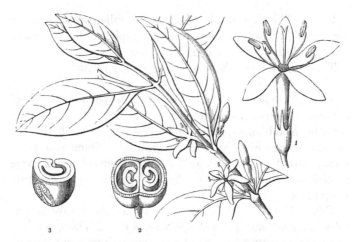

Coffea Arabica. 1. A flower magnified. 2. A section across a ripe fruit. 3. A portion of a seed, showing the small embryo laid bare in the end of convolute albumen.

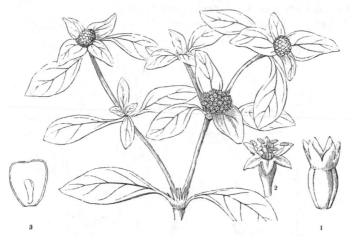

Richardsonia scabra. 1. An ovary with its calyx. 2. A corolla. 3. A vertical section of a seed, with an erect embryo in copious albumen.

87. — *Dipsaceæ.* Herbaceous plants or under-shrubs. Leaves opposite or whorled. Flowers capitate, surrounded by a many-leaved involucre. Calyx superior, membranous; surrounded by an involucel. Corolla oblique, imbricated. Stamens 4; anthers distinct. Ovary 1-celled, with a pendulous ovule; stigma simple. Fruit crowned by the pappus-like calyx, embryo in fleshy albumen.

Uses.—The ripe heads of Dipsacus fullonum, dried, are formed of hard stiff spines, and are employed by fullers, in dressing cloth, under the name of teasels.

Typical Genera.—Scabiosa, Dipsacus, Knautia.

88. — *Valerianaceæ.* Herbs. Leaves opposite, without stipules. Flowers corymbose, panicled, or in heads. Calyx superior, membranous, or resembling pappus. Corolla tubular, regular or irregular, sometimes calcarate. Stamens 1 to 5. Ovary with 1 perfect cell, and 2 other abortive ones; ovule pendulous; stigmas 1 to 3. Fruit dry. Embryo destitute of albumen.

Uses.—Common Valerian, and several others, have powerfully aromatic, antispasmodic, febrifugal roots. The genus Valerianella consists of annual herbs, whose leaves are used as salad, under the name of Lamb's lettuce. The Spikenard of the ancients was Nardostachys Jatamansi.

Typical Genera.—Valerianella, Centranthus, Valeriana.

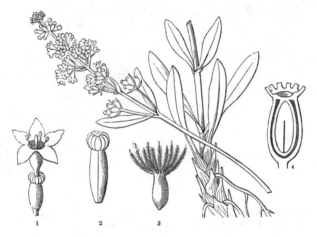

Valeriana Celtica. 1. An entire flower magnified. 2. The ovary and young calyx. 3. The fruit, with the pappose full-grown calyx. 4. A vertical section of a ripe fruit and seed.

89. — *Asteraceæ* or *Compositæ.* Shrubs, or herbs, extremely variable in appearance. Flowers in heads, surrounded by an involucrum, and seated on a receptacle, from which paleæ often spring. Calyx obsolete; a pappus (328). Co-

rolla regular or irregular. Anthers united into a tube.
Ovary inferior, one-celled, with an erect ovule. Embryo without albumen.

Division 1.—*Cichoraceæ.* Florets all ligulate. Milky.
TYPICAL GENERA.—Hieracium, Taraxacum.

Division 2.—*Corymbiferæ.* Florets in part or wholly tubular. Corolla funnel-shaped. Involucrum hemispherical, leafy or scaly, soft, seldom spiny.
TYPICAL GENERA.—Chrysanthemum, Tussilago.

Division 3.—*Cynaraceæ.* Florets wholly tubular. Corolla with a ventricose throat. Involucrum hard, conical, and generally spiny.
TYPICAL GENERA.—Carduus, Cynara, Onopordum.

Division 4.—*Labiatifloræ.* Florets bilabiate.
TYPICAL GENERA.—Mutisia, Triptilion.

USES.—Among the *Cichoraceous* division a narcotic principle is commonly found, which in the garden Lettuce is so diffused as to be bland, and in Lactuca virosa is so concentrated as to render the extract similar to opium in effect. Succory, Endive, Salsafy, Scorzonera, well-known esculents, belong here. Of the *Corymbiferous* division, Chamomile is characteristic, with its bitter tonic qualities. Many others, such as Coltsfoot, Elecampane, Feverfew, correspond in properties with Chamomile. Wormwood, Southernwood, species of Artemisia, are aromatic and extremely bitter; Tarragon, a pungent herb, used for pickling, is Artemisia Dracunculus. Pellitory of Spain, which is acrid, and excites the salivary organs powerfully, is Anacyclus Pyrethrum; and similar effects are produced by Spilanthus oleracea, Bidens tripartita, and others. The Sunflower, Guizotia oleifera, Madia sativa, and others, yield a bland oil when their seeds are pressed. Jerusalem artichokes, a well-known article of food, are the tubers of Helianthus tuberosus. The *Cynaraceous* division consists principally of bitter plants. Centaurea calcitrapa, Silybum (or Carduus) Marianum, Cnicus Benedictus, and the common Burdock, are all stomachics of some importance. The flowers of Carthamus tinctorius are dried for the use of the dyers, and resemble Saffron. The fleshy receptacles of Cynara Scolymus are the artichoke bottoms of our kitchens.

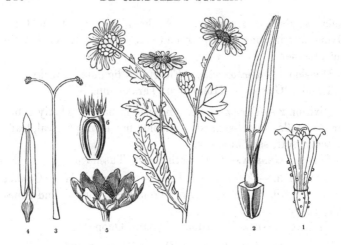

Argyranthemum Jacobæifolium.   1. A tubular floret of the disk.   2. A ligulate floret of the ray.   3. Style and stigmas.   4. An anther.   5. An involucrum and conical receptacle, from which the florets have fallen.   6. Ripe achænium cut through vertically, with toothed coronetted pappus.

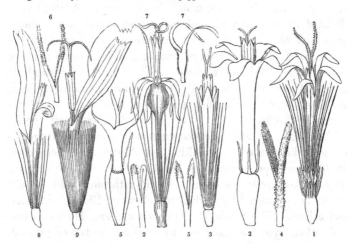

The old divisions of this large order are adhered to because they appear more likely to be permanent than the more recent suborders, &c. proposed by De Candolle and others, in which peculiarities in the stigma are chiefly employed. The student who desires to become acquainted with the details of this enormous order, numbering more than 7000 species, will consult De Candolle's Prodromus, Vols. V. VI.

and VII. The preceding wood-cut will assist him in under-
standing the distinctions of that author.

1. Tubular floret of Webbia aristata, with double pappus (*Vernoniaceæ*, D.C.) 2.
Tubular floret and stigma of Anisochæta mikanioides, with pappus of 4 setæ (*Eupa-
toriaceæ*, D.C.) 3. Tubular floret of Berthelotia lanceolata, with silky pappus (*Asteroideæ*, D.C.) 4. Stigma of Blumea senecioides (*Asteroideæ*, D.C.) 5. Ligulate floret
and stigma of Lipochæta umbellata ; pappus of two unequal winged paleæ (*Senecionideæ*, D.C.) 6. Stigma of Dunantia achyranthes (*Senecionideæ*, D.C.) 7. Tubular
floret with ventricose throat and the stigma of Aplotaxis Nepalensis (*Cynareæ*, D.C.)
8. Ligulate bilabiate floret of Oreoseris lanuginosa (*Mutisiaceæ*, D.C.) 9. Ligulate
floret of Brachyramphus obtusus (*Cichoraceæ*, D.C.)

90. — *Galiaceæ*, or *Stellatæ*. Herbaceous plants, with
whorled leaves, destitute of stipules. Stems usually angular.
Calyx 4- 5- or 6-lobed. Corolla valvate, rotate or tubular,
regular. Stamens equal in number to the lobes of the corolla,
and alternate with them. Ovary 2-celled ; ovules solitary,
erect. Fruit a didymous, indehiscent pericarp. Embryo
minute, straight, in horny albumen.

USES.—The roots of Rubia tinctorum yield madder, a
quality in which others participate, though in a less degree.
The yellow flowers of Galium verum are used to curdle milk.
The fragrant Woodruff has the reputation of being diuretic ;
Asperula cynanchica is said to be astringent. Except the
species used for dyeing, none are of any real importance.

TYPICAL GENERA.—Asperula, Galium, Rubia.

91.—*Goodeniaceæ*. Herbaceous plants, rarely shrubs, without milk. Leaves scattered, without stipules. Flowers
showy. Calyx superior, equal or unequal. Corolla more or
less irregular, split at the back ; the segments folded inwards
in æstivation. Stamens 5, distinct. Ovary with indefinite
ovules ; stigma surrounded by a membranous cup. Fruit
a capsule ; albumen fleshy.

USES.—Unknown. TYPICAL GENERA.—Goodenia, Euthales.

92. — *Scævolaceæ*. Herbaceous plants with the flowers
axillary or terminal, and never in heads. Calyx superior,
sometimes obsolete. Corolla irregular, split at the back,
the edges of the divisions folded inwards in æstivation. Stamens 5, distinct ; anthers distinct or united. Ovary few-
celled, with solitary erect ovules ; stigma surrounded by a
cup. Fruit drupaceous or nucamentaceous.

USES.—Unknown.

TYPICAL GENERA.—Scævola, Dampiera.

93.— *Stylidiaceæ.*   Glandular herbs.   Calyx superior, 2-6-parted, permanent.   Corolla irregular, imbricated.   Stamens 2, connate into an elastic slender column, with which the style is consolidated.   Ovary 2-celled.   Capsule 2-valved, many-seeded.   Seeds albuminous.

Uses.—Unknown.   Remarkable for the irritable elastic column of stamens.

Typical Genera.—Stylidium, Leuwenhoekia.

94.—*Lobeliaceæ.*   Herbaceous milky plants or shrubs. Leaves alternate, without stipules.   Flowers usually showy. Calyx superior, 5-lobed, or entire.   Corolla irregular, 5-lobed, or 5-cleft.   Stamens 5 ; anthers cohering.   Stigma fringed. Fruit capsular, 1 or more celled, many-seeded ; embryo in the axis of albumen.

Uses.—The species abound in a milky juice of extreme acridity.   Lobelia inflata is an emetic, but dangerous from its violence.   Hippobroma (or Isotoma) longiflora is fatally hypercathartic.   Many are plants of great beauty.

Typical Genera.—Lobelia, Clintonia, Siphocampylus.

95.—*Gesneraceæ.*   Herbaceous plants or under-shrubs. Leaves opposite, rugose, fleshy, without stipules.   Flowers generally showy.   Calyx half superior, valvate.   Corolla tubular, with an imbricate æstivation.   Anthers cohering, with a thick connective.   Ovary 1-celled, surrounded by glands, with 2-lobed polyspermous placentæ ; stigma capitate.   Embryo in the axis of albumen.

Uses.—The succulent fruit is eatable.   Some species yield a dyeing substance.   The species are, however, of no real importance ; but they are generally gay flowers.

Typical Genera.—Gloxinia, Gesnera, Columnea.

96.—*Campanulaceæ.*   Herbaceous plants or under-shrubs, yielding a white milk.   Leaves alternate, without stipules. Flowers usually showy.   Calyx superior, permanent.   Corolla usually 5-lobed, withering, regular, valvate.   Stamens alternate with the lobes of the corolla.   Anthers distinct. Style covered by collecting retractile hairs ; stigma naked. Fruit dehiscing by apertures, or valves.   Seeds numerous ; embryo in the axis of albumen.

Uses.—Slightly acrid, but not dangerous. Rampion, a root used like Radishes, is Campanula Rapunculus.

Typical Genera.—Campanula, Phyteuma, Roella.

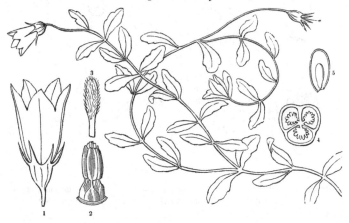

Wahlenbergia procumbens. 1. An entire flower. 2. Stamens. 3. A stigma. 4. Transverse section of the ovary. 5. A vertical section of a seed, showing the embryo.

97.—*Vaccinaceæ.* Are the same as Ericaceæ, only the ovary is inferior.

Uses.—The bark is slightly astringent, and fruit succulent. Cranberries are the fruit of species of Oxycoccus, Bilberries and Whortleberries of species of Vaccinium.

Typical Genera.—Vaccinium, Thibaudia.

98.—*Ericaceæ.* Shrubs or under-shrubs. Leaves evergreen, rigid, without stipules. Calyx 4- or 5-cleft, inferior. Corolla hypogynous, 4- or 5-cleft, imbricated. Stamens definite, hypogynous; anthers 2-celled, dehiscing by a pore. Ovary many-celled, many-seeded; style 1. Fruit capsular. Seeds indefinite, minute; embryo in the axis of albumen.

Uses. — Loiseleuria procumbens, Rhododendron ferrugineum, and others, are astringent. Arctostaphylus Uva Ursi is diuretic. Rhododendron Chrysanthum is a powerful narcotic, and this seems to be a general character of the order, some of which, as Rhododendron maximum, Kalmia latifolia, and Azalea Pontica, are dangerous poisons. Most of the species are plants of great beauty.

Typical Genera.—Rhododendron, Kalmia, Erica.

Rhododendron albiflorum. 1. A corolla and pistil, with all the stamens removed save one. 2. An anther. 3. A ripe capsule burst. 4. A vertical section of a seed.

99.—*Ebenaceæ*. Trees or shrubs without milk. Leaves alternate, coriaceous. Calyx inferior, in 3 or 6 divisions. Corolla hypogynous, usually pubescent, imbricated. Stamens definite; twice as many as the segments of the corolla, four times as many, or the same number. Ovary several-celled, the cells having 1 or 2 pendulous ovules; style divided. Fruit fleshy, few-seeded. Albumen cartilaginous; embryo in the axis; radicle turned towards the hilum.

Subdivision.—*Styraceæ*. Ovary inferior. Stamens perigynous. Style simple.

Uses.—The fruit of Diospyros Lotus, Kaki, and others, is extremely astringent when green, but becomes bletted and sweet after a time, when it is eaten. Diospyros Virginiana and others have a febrifugal bark. Ebony is the wood of Diospyros Ebenus and several other species of that genus. The fragrant gum resins, Storax and Benzoin, are produced by species of Styrax.

Typical Genera.—Diospyros, Maba, Styrax.

100.—*Aquifoliaceæ*. Trees or shrubs. Leaves coriaceous. Flowers small. Sepals inferior, 4 to 6, imbricated. Corolla hypogynous. Stamens alternate with its segments. Disk none. Ovary with from 2 to 6 cells; ovules solitary, pendulous. Fruit indehiscent, with from 2 to 6 stones. Seed suspended; albumen large; embryo small, 2-lobed.

USES. — Ilex Aquifolium, the common Holly, has leaves emetic, and berries purgative; its leaves are powerfully febrifugal. The fruit and bark of Prinos verticillatus and others have similar qualities. Paraguay tea is Ilex Paraguensis. Some are diuretic.

TYPICAL GENERA.—Ilex, Prinos, Cassine.

101.—*Sapotaceæ.* Trees or shrubs with milky juice. Leaves alternate, without stipules, coriaceous. Calyx inferior, regular, permanent. Corolla hypogynous; its segments usually equal in number to those of the calyx, seldom twice or thrice as many. Stamens arising from the corolla, definite. Anthers usually turned outward; sterile stamens usually present. Ovary with several cells, and one erect ovule in each cell. Style 1. Fruit baccate. Seeds nut-like. Testa bony, shining. Embryo large, usually in fleshy albumen.

USES.—The species are generally astringent and febrifugal. Achras Sapota and others are the Sapodilla plums, whose fruit is much esteemed in the West Indies. The Star-apple (another West Indian fruit) is Chrysophyllum Cainito. A vegetable butter is yielded by some species of Bassia.

TYPICAL GENERA.—Achras, Chrysophyllum, Mimusops.

102.—*Myrsinaceæ.* Trees or shrubs. Leaves alternate, serrated, coriaceous; stipules wanting. Calyx 4- or 5-cleft. Corolla hypogynous. Stamens opposite the segments of the corolla; sometimes 5 sterile, petaloid, additional filaments. Ovary 1, with a free central placenta; style 1. Fruit fleshy, mostly 1-seeded. Seeds peltate, albumen horny; embryo lying across the hilum.

USES.—Embelia robusta is said to have purgative berries.

TYPICAL GENERA.—Ardisia, Myrsine.

103.—*Oleaceæ.* Trees or shrubs. Branches usually dichotomous. Leaves opposite. Calyx monophyllous, permanent. Corolla hypogynous, 4-cleft, valvate. Stamens 2. Ovary without any disk, 2-celled; ovules pendulous; stigma bifid or undivided. Fruit often 1-seeded. Seeds with dense albumen.

USES.—Olive oil is obtained from the fruit of Olea europæa. Manna exudes from the trunk of Ornus europæa and others. The bark of the Olive is a powerful febrifuge. Phyllireas are handsome evergreen shrubs.

TYPICAL GENERA.—Olea, Phyllirea, Syringa.

104.—*Jasminaceæ.* Shrubs. Leaves opposite or alternate, mostly compound. Calyx divided or toothed, permanent. Corolla regular, with from 5 to 8 divisions, imbricated and twisted. Stamens 2. Ovary destitute of a disk, 2-celled; ovules erect; stigma 2-lobed. Seeds with no albumen.

USES.—The flowers of most species are fragrant. The leaves and bark are bitter, but of little moment.

TYPICAL GENERA.—Jasminum, Nyctanthes.

105.—*Apocynaceæ.* Trees or shrubs, usually milky. Leaves opposite, quite entire, often having glands upon the petioles, with no stipules. Calyx inferior, permanent. Corolla regular, 5-lobed, contorted. Stamens 5. Filaments distinct. Pollen granular. Ovaries 2, or 1 which is 2-celled, polyspermous. Stigma 1. Seeds with fleshy albumen.

USES.—Often dangerous poisons, but in some cases simply purgatives. The root of Nerium, the kernel of Tanghinia venenata, the seeds of various kinds of Strychnos, called Nux vomica, belong to the first class; the leaves of Cerbera Manghas, Allamanda cathartica, to the second. Vahea, Urceola elastica, and others, abound in Caoutchouc.

TYPICAL GENERA.—Vinca, Echites, Nerium.

106.—*Asclepiadaceæ.* Shrubs or herbaceous plants, milky, and often twining. Leaves entire, opposite, having ciliæ between their petioles. Calyx inferior, permanent. Corolla 5-lobed, regular, imbricated, very seldom valvular. Stamens 5; filaments connate; anthers 2-celled; pollen cohering in masses, and sticking by 5 glands to as many processes of the stigma. Ovaries 2. Styles 2. Stigma common to both styles, 5-cornered. Follicles 2. Seeds comose; albumen thin.

USES.—The roots of many are emetic, sudorific, acrid, and purgative. Indian Sarsaparilla is the root of Hemidesmus indicus. Asclepias tuberosa and Curassavica are employed as cathartics in the United States and West Indies. The leaves of Cynanchum Argel are used in Egypt to adulterate Senna; they are acrid. The extract of Calotropis gigantea, the Mudar plant, is powerfully alterative and purgative. Many species have a tough fibre, which renders them fit for cordage; others yield abundance of Caoutchouc.

TYPICAL GENERA.—Periploca, Stapelia, Physianthus.

Schubertia multiflora. 1. The anthers united to the stigma. 2. The ovary and stigma, from the latter of which the pollen masses have been removed. 3. A pair of pendulous pollen masses, with their gland. 4. The ripe follicles.

107.—*Bignoniaceæ.* Trees or shrubs, often twining or climbing. Leaves opposite, usually compound, without stipules. Flowers large and showy. Calyx inferior, sometimes spathaceous. Corolla irregular. Stamens 5, of which 1 always and sometimes 3 are sterile. Ovary in a disk, 2-celled, polyspermous; style 1; stigma of 2 plates. Fruit berried or capsular; if the latter, 2-valved, 2-celled, long and compressed. Seeds often winged; albumen 0.

USES.—Usually beautiful plants. Some have hard timber, and a red fecula is obtained from the leaves of Bignonia Cherere and others. The genera with berried fruit form a peculiar division, and include Crescentia Cujete, the Calabash-tree, and Parmentiera edulis, both of which have eatable fruit.

TYPICAL GENERA.—Bignonia, Tecoma.

108.—*Cyrtandraceæ.* Herbs. Leaves opposite, often radical. Flowers showy. Calyx inferior, campanulate, equal. Corolla irregular, imbricated. Stamens didynamous. Disk

annular.   Ovary 1-celled, with 2 double placentæ ; stigma
2-lobed.   Fruit capsular and siliquose, or succulent, many-
seeded.   Seeds minute, often with tails ; albumen absent.
Uses.—Unknown.
Typical Genera.—Æschynanthus, Streptocarpus.

109.—*Gentianaceæ.*   Herbaceous plants.   Leaves opposite,
entire, without stipules, usually 3-5-ribbed.   Flowers showy.
Calyx inferior, permanent.   Corolla regular, with an imbricated, twisted, or plaited æstivation.   Stamens inserted upon
the corolla, some of them occasionally abortive.   Ovary 1-
celled ; stigmas 1 or 2.   Capsule or berry many-seeded ; the
margins of the valves turned inwards.   Embryo in the axis of
soft albumen.

Uses.—All the species are more or less bitter ; many in-
tensely so.   The Gentian root of the shops is obtained from
Gentiana lutea chiefly ; the leaves and stems of Agathotes
Chirayta furnish the Gentian of India.   Menyanthes trifoliata
is the Buck-bean, employed advantageously as a tonic.
Typical Genera.—Erythræa, Gentiana, Chironia.

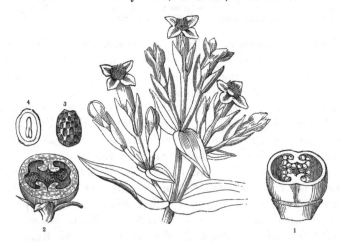

Gentiana amarella. 1. Section of the ovary of Chironia baccifera. 2. Section of
the ripe fruit. 3. A seed. 4. A vertical section of it.

110.—*Polemoniaceæ.*   Herbaceous plants.   Leaves opposite.   Calyx inferior, 5-parted.   Corolla regular, 5-lobed.

Stamens 5, unequal, on the tube of the corolla. Ovary 3-celled; stigma 3-lobed. Capsule 3-celled; 3-valved, the valves separating from the axis. Embryo in horny albumen.

Uses.—Unknown.

Typical Genera.—Polemonium, Phlox, Gilia.

111.—*Convolvulaceæ.* Herbaceous plants, or shrubs, usually twining and milky. Leaves alternate. Calyx permanent, inferior, in 5 divisions, remarkably imbricated, often unequal. Corolla hypogynous, plaited. Stamens 5, inserted towards the base of the corolla. Ovary with 2 to 4 cells, few seeded; ovules erect; style 1. Disk annular. Capsule with the valves fitting at their edges to the angles of a loose dissepiment. Seeds with mucilaginous albumen; embryo curved; cotyledons shrivelled.

Uses.—The roots of Convolvulus Scammonia yield Scammony; of Exogonium Purga, true Jalap; of Ipomœa Batatoides, a kind of false Jalap, called Purga Macho; and a great many more possess similar properties. The Batatas, or Sweet Potatoe, has the purgative quality so much diffused as to be a valuable article of food; the great roots of others have also been found eatable.

Typical Genera.—Ipomœa, Convolvulus, Calystegia.

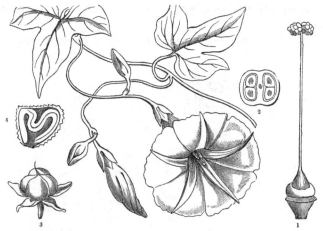

Ipomœa Batatoides. 1. The pistil and annular disk. 2. A transverse section of the ovary. 3. A capsule of Convolvulus tricolor. 4. A vertical section of the seed of that species.

112.—*Cuscutaceæ.* Leafless parasites. Calyx permanent, inferior, 4-5-parted, imbricated. Corolla permanent, imbricated. Scales alternating with segments of corolla. Stamens opposite the last. Ovary 2-celled; ovules in pairs, erect; styles 2. Capsule. Embryo spiral, in fleshy albumen.

USES.—Unknown.

TYPICAL GENUS.—Cuscuta.

113.—*Cordiaceæ.* Trees. Leaves scabrous, without stipules. Calyx inferior, 5-toothed. Corolla regular. Stamens alternate with the segments of the corolla. Ovary 4-celled, with 1 pendulous ovule in each cell; stigma 4-cleft. Fruit drupaceous, 4-celled. Seed pendulous by a funiculus; cotyledons plaited; albumen 0.

USES.—Unimportant. Sebesten plums, an emollient mucilaginous fruit, are produced by Cordia Myxa, and Sebestena.

TYPICAL GENUS.—Cordia.

114.—*Boraginaceæ.* Herbaceous plants, or shrubs. Stems round. Leaves alternate, covered with asperities. Flowers in gyrate racemes (scorpioid). Calyx inferior, permanent. Corolla hypogynous, regular. Stamens 5, inserted upon the corolla. Ovary 4-parted, 4-seeded; style simple; stigma simple or bifid. Nuts 4, distinct. Seed without albumen.

USES.—The dye called Alkanet is obtained from the roots of Anchusa tinctoria and several other species. The foliage is insipid and harmless.

TYPICAL GENERA.—Myosotis, Anchusa, Lithospermum.

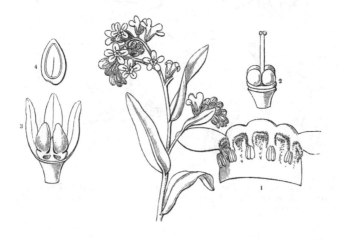

Myosotis. 1. Throat cut open. 2. A pistil. 3. Ripe fruit with two of the nuts remaining, and the scars of two that have dropped off. 4. A perpendicular section of a nut.

115.—*Solanaceæ.* Herbaceous plants or shrubs. Leaves alternate, sometimes collateral. Inflorescence often out of the axil; pedicels without bracts. Calyx permanent, inferior. Corolla regular, or somewhat unequal, plaited. Stamens inserted upon the corolla. Ovary 2-celled; stigma simple. Pericarp with 2 or 4 cells. Seeds numerous; embryo usually curved in fleshy albumen.

USES.—Many are narcotic, as Tobacco, Henbane, Stramonium, Bitter-sweet, and Deadly Nightshade, or Belladonna. The fruit of others is almost free from deleterious qualities, and eatable; as the Aubergine, Solanum esculentum, Tomatoes, or Solanum Lycopersicon, Physalis edulis, and many others. In some species starch is collected in great quantity, and renders them fit for food, as in the tubers of the Potatoe, Solanum tuberosum.

TYPICAL GENERA.—Solanum, Datura, Physalis.

Petunia violacea. 1. A cross section of the ovary. 2. Ripe fruit of Solanum Dulcamara. 3. A section of one of its seeds.

116. — *Hydrophyllaceæ.* Herbaceous plants. Leaves usually lobed. Inflorescence often gyrate. Calyx inferior, 5-cleft, with reflexed appendages. Corolla regular. Stamens 5, epipetalous. Ovary simple, 1-celled; placentæ 2. Fruit 2-valved. Seeds reticulated; embryo cartilaginous.

Uses.—Unknown.

Typical Genera.—Nemophila, Phacelia.

117.—*Orobanchaceæ*. Parasitical brown leafless herbs.
Calyx permanent. Corolla irregular. Stamens didynamous.
Ovary 1-celled, in a fleshy disk, with 2 or more parietal placentæ ; stigma 2-lobed. Fruit capsular, many-seeded, enclosed within the withered permanent corolla ; seeds very
minute ; embryo extremely small, in the apex of albumen.

Uses.—Scarcely known ; they are astringent plants.

Typical Genera.—Orobanche, Lathræa.

118.—*Scrophulariaceæ*. Herbs or shrubs with opposite or
alternate exstipulate leaves. Calyx tubular, permanent. Corolla irregular. Stamens didynamous, or 2. Ovary 2-celled;
ovules numerous ; stigma 2-lobed. Fruit 2-celled ; seeds indefinite or definite, albuminous.

Uses.—Foxglove, whose action upon the pulse is so lower-
ing, is Digitalis purpurea. Gratiola officinalis, some Calceo-
larias, and others, are purgative and emetic. Euphrasia
officinalis is bitter and sub-aromatic. Vandellia diffusa is a
powerful antibilious emetic and febrifuge.

Typical Genera.—Scrophularia, Antirrhinum, Pentstemon.

Digitalis purpurea. 1. A corolla split open. 2. A pistil. 3. A transverse section of it. 4. A ripe capsule. 5. A vertical section of a seed.

119.—*Lamiaceæ* or *Labiatæ.* Herbaceous plants or under-shrubs. Stem 4-cornered. Leaves opposite, often replete with aromatic oil. Flowers in axillary cymes; sometimes solitary. Calyx tubular, permanent. Corolla bilabiate. Stamens didynamous, the 2 upper sometimes wanting. Ovary 4-lobed; style 1; stigma bifid. Fruit 1 to 4 small nuts. Seeds with little or no albumen.

Uses.—The species are always harmless, and in many cases useful for their tonic aromatic qualities. Lavender is Lavandula vera; Horehound, used for coughs, is Marrubium vulgare. Savory, Mint, Marjoram, Thyme, Sage, are all pot-herbs used in cookery. Teucrium Marum is a powerful and singular stimulant of cats. The cordial Peppermint is prepared from Mentha piperita. A kind of stearoptene resembling Camphor, is found in many species.

Typical Genera.—Lamium, Salvia, Scutellaria.

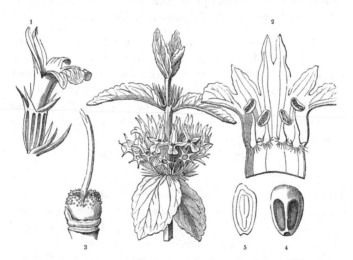

Marrubium vulgare. 1. An entire flower seen in profile. 2. A corolla slit open. 3. The pistil. 4. A nut. 5. A vertical section of the latter, showing the embryo.

120.—*Verbenaceæ.* Trees or shrubs, sometimes herbaceous plants. Leaves opposite, without stipules. Flowers in opposite corymbs, or spiked alternately; sometimes in dense heads. Calyx tubular. Corolla irregular. Stamens didynamous, occasionally 2. Ovary 2- or 4-celled; ovules erect or pendulous; style 1; stigma bifid. Fruit composed of 2 or 4 nucules in a state of adhesion; albumen none.

Uses.—A few are slightly aromatic and bitter.

Typical Genera.—Verbena, Aloysia, Callicarpa.

121.—*Acanthaceæ.* Herbaceous plants or shrubs. Leaves opposite, without stipules. Inflorescence in spikes, racemes, fascicles, or even solitary. Flowers usually opposite, placed within bracts. Calyx very much imbricated, permanent, inferior. Corolla irregular, 2-lipped. Stamens mostly 2, sometimes didynamous. Ovary in a disk, 2-celled, 2- or many-seeded; stigma 2-lobed. Capsule 2-celled, bursting elastically. Seeds hanging by hard, usually hooked processes of the placentæ; albumen none.

Uses.—Acanthus spinosus is accounted emollient. The leaves and roots of Adhatoda Vasica are supposed to be antispasmodic. Justicia paniculata is bitter and stomachic.

Typical Genera.—Ruellia, Justicia, Eranthemum.

122. — *Lentibulaceæ.* Herbaceous plants. Leaves undivided, or resembling roots, and bearing vesicles. Flowers single, or in spikes. Calyx permanent, inferior. Corolla irregular, bilabiate, with a spur. Stamens 2; anthers simple. Ovary 1-celled, with a free central placenta; stigma bilabiate. Capsule 1-celled. Seeds without albumen.

Uses.—Of no importance.

Typical Genera.—Pinguicula, Utricularia.

123. — *Plumbaginaceæ.* Herbaceous plants or shrubs. Leaves alternate, undivided, somewhat sheathing. Calyx inferior, tubular, plaited. Corolla regular. Stamens definite. Ovary superior, 1-seeded; ovule pendulous from an umbilical cord; styles 5. Fruit a utricle. Seed inverted.

Uses.—Statice Limonium, and others, have extremely astringent roots. The bark of Plumbago is acrid and vesicant.

Typical Genera.—Armeria, Plumbago.

124.—*Globulariaceæ.* Shrubs or herbs. Leaves alternate. Flowers in heads. Calyx inferior, permanent, 5-cleft, sometimes 2-lipped. Corolla hypogynous, bilabiate, made up of

5 parts. Stamens 4, from the tube of the corolla. Ovary superior, 1-celled, with a pendulous ovule. Albumen fleshy.

Uses.—The species are said to be bitter, tonic, and purgative ; they appear to be of little importance.

Typical Genus.—Globularia.

125.—*Primulaceæ.* Herbaceous plants. Calyx 4-5-cleft, permanent, inferior. Corolla regular. Stamens inserted upon the corolla opposite its segments. Ovary 1-celled, with a free central placenta ; style 1 ; stigma capitate. Capsule with a central placenta. Embryo lying across the hilum in fleshy albumen.

Uses.—The root of Cyclamen is acrid ; the flowers of Cowslips sedative. Anagallis arvensis is powerfully acrid.

Typical Genera.—Primula, Anagallis, Lysimachia.

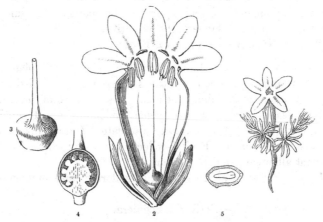

1. Aretia Vitaliana.   2. A flower cut open.   3. The pistil.   4. A vertical section of the latter, showing the free central placenta.   5. A section of a seed.

126. — *Plantaginaceæ.* Herbaceous plants, with spiked inconspicuous flowers, and ribbed leaves. Calyx inferior, 4-leaved, imbricated. Corolla membranous, hypogynous, 4-parted. Stamens 4 ; filaments flaccid ; anthers versatile. Ovary without a disk ; ovules peltate or erect, solitary, twin, or indefinite ; stigma hispid, simple. Capsule membranous. Embryo in fleshy albumen.

Uses.—The species are of little importance. The seeds of

Plantago Psyllium and others are mucilaginous; the foliage of Pl. angustifolia is slightly astringent; this plant, which is commonly called Plantain or Ribgrass, is of some value for sheep-feed in dry exposed places.

TYPICAL GENERA.—Plantago, Littorella.

SUBCLASS IV. MONOCHLAMYDEÆ.

127.—*Phytolaccaceæ.* Under-shrubs or herbaceous plants. Leaves alternate, without stipules, often with pellucid dots. Calyx inferior, of 4 or 5 petaloid leaves. Stamens indefinite, or, if equal to the number of the divisions of the calyx, alternate with them. Ovary of from 1 to several cells, each containing 1 ascending ovule. Fruit baccate or dry, 1- or many-celled. Seeds solitary, with a cylindrical embryo curved round mealy albumen.

USES.—The succulent fruit of Phytolacca decandra is said to be useful in chronic and siphylitic rheumatism; its juice is acrid, emetic, and dangerously purgative.

TYPICAL GENERA.—Phytolacca, Rivina.

128.—*Petiveriaceæ.* Under-shrubs or herbaceous plants, with an alliaceous odour. Leaves alternate, with distinct stipules, often with minute pellucid dots. Calyx of several distinct leaves. Stamens perigynous, indefinite, or, if equal to the segments of the calyx, alternate. Ovary superior, 1-celled; ovule erect. Fruit 1-celled, indehiscent, dry. Seed without albumen; radicle inferior.

USES.—Petiveria alliacea is acrid, sudorific, and emmenagogue.

TYPICAL GENERA.—Petiveria, Seguiera.

129.—*Chenopodiaceæ.* Herbaceous plants or under-shrubs. Leaves alternate without stipules. Flowers small. Calyx sometimes tubular at the base, persistent. Stamens inserted into the base of the calyx, opposite its segments. Ovary superior, with a single ovule attached to the base of the cavity. Fruit membranous. Embryo curved round farinaceous albumen, or spiral, or doubled up without albumen.

USES.—Spinach, Garden Orach (Atriplex hortensis), Chard Beet, and Sea Beet, are delicate esculents whose leaves are eaten boiled. The roots of common Beet and Mangel Wurzel are succulent, sweet, and valuable for food. The seeds of Chenopodium Quinoa are extensively consumed for food in Peru. On the other hand, Chenopodium olidum and baryosmon are fœtid emmenagogues ; Ch. anthelminticum furnishes the anthelmintic oil of wormseed ; Ch. ambrosioides is a fragrant expectorant. Several species of Atriplex are reported to have emetic seeds. Various kinds of Salsola and Salicornia supply the sodas of the shops.

TYPICAL GENERA.—Chenopodium, Atriplex, Blitum.

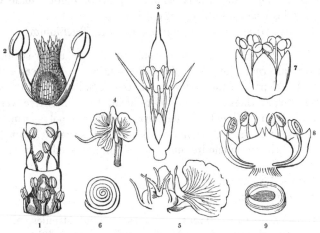

1. A portion of the spike of Salicornia herbacea, with the flowers lodged in the notches of the axis. 2. A flower separate. 3. A flower of Salsola Kali. 4. Its ripe fruit. 5. The same magnified, with a portion of the leafy dilated calyx torn away. 6. Its embryo. 7. A flower of Chenopodium album. 8. A section of the same, showing the superior ovary. 9. Its seed cut through to show the embryo.

130.—*Nyctaginaceæ*. Stem either herbaceous, shrubby, or arborescent. Leaves opposite, and almost always unequal ; sometimes alternate. Flowers having either a common or proper involucre. Calyx tubular, sometimes coloured ; becoming indurated at the base. Stamens definite, hypogynous. Ovary superior, with a single erect ovule. Fruit a utricle, enclosed within the base of the calyx. Embryo with foliaceous cotyledons, wrapping round floury albumen.

USES.—The fleshy roots of the species of Mirabilis are slightly purgative.

TYPICAL GENERA.—Mirabilis, Oxybaphus.

131.—*Amarantaceæ.* Herbs or shrubs. Leaves simple, without stipules. Flowers in heads or spikes, usually coloured. Calyx scarious, persistent, immersed in dry coloured bracts. Stamens hypogynous. Ovary superior, 1- or few-seeded; ovules hanging from a free central funiculus. Fruit a utricle. Seeds lentiform; albumen farinaceous; embryo curved round the circumference; radicle next the hilum.

USES.—Unimportant. The species are insipid, on which account some species of Amaranthus have been employed as spinach. Their dry richly coloured flowers render some of the species beautiful objects of cultivation.

TYPICAL GENERA.—Amaranthus, Celosia, Trichinium.

132.—*Begoniaceæ.* Herbaceous plants or under-shrubs. Leaves alternate, oblique. Stipules scarious. Flowers unisexual. Sepals in the males 4; in the females 5. Stamens indefinite; anthers collected in a head, the connective very thick. Ovary winged, 3-celled, with 3 double polyspermous placentæ in the axis; stigmas 3, somewhat spiral. Fruit 3-celled, with an indefinite number of minute seeds; embryo without albumen.

USES.—Unknown.

TYPICAL GENUS.—Begonia.

133.—*Lauraceæ.* Trees. Leaves without stipules, alternate. Calyx 4-6-cleft, imbricated. Stamens definite, perigynous; anthers 2-4-celled, bursting by recurved valves. Glands at the base of the inner filaments. Ovary superior, with one or two pendulous ovules. Fruit fleshy. Seed without albumen; embryo amygdaloid, with peltate cotyledons.

USES.—All appear to be aromatics, although some, as Oreodaphne fœtens and others, have the aromatic principle so concentrated as to be acrid. The seeds of Nectandra Puchury and Aydendron Cujumary are the Pichurim beans or Sassafras nuts, used as a substitute for nutmegs. Cinnamomum zeylanicum yields cinnamon, and a bark of like nature is supplied by many other plants of this order. Camphor is obtained from Camphora officinarum; and the aromatic Sassafras bark, used by the people of the United States as a powerful sudorific, is taken from the root of Sassafras officinale. The

Avocado pear, an eatable West Indian fruit, is borne by Persea gratissima.

TYPICAL GENERA.—Laurus, Cinnamomum.

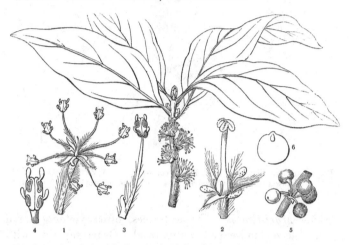

Litsæa Baueri. 1. A male flower. 2. A female. 3. A stamen, with a gland at the base. 4. An anther, with the recurved valves. 5. A cluster of fruit. 6. A cotyledon seen from within, with the plumula adhering to the inner face.

134.—*Polygonaceæ*. Herbaceous plants, rarely shrubs. Leaves alternate, their stipules usually cohering in the form of an ochrea. Calyx inferior, imbricated. Stamens definite. Ovary superior, with a single erect ovule. Nut triangular. Seed with farinaceous albumen; embryo inverted; radicle remote from the hilum.

USES.—Rumex scutatus, Acetosa, and others, are the Sorrel plants used in cookery. Rhubarb is the root of several species of Rheum; similar properties, only more feeble, are found in Rumex alpinus. In addition to acid and purgative qualities, a great degree of astringency manifests itself, as in the roots of Rumex, and the bark of Coccoloba uvifera, which is said to yield a kind of Kino. It is reported that the seeds of Polygonum aviculare are emetic, notwithstanding that those of P. Fagopyrum and tataricum are employed as food in some places; the leaves of Polygonum hydropiper and others are acrid.

TYPICAL GENERA.—Rumex, Polygonum, Rheum.

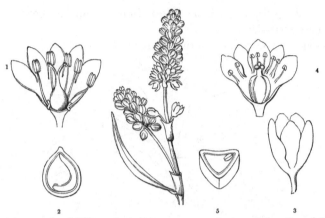

Polygonum lapathifolium. 1. A flower cut open. 2. A vertical section of the seed. 3. A flower of Polyg. Convolvuli. 4. The same cut open. 5. A transverse section of a seed.

135.—*Myristicaceæ.* Tropical trees, often yielding a red juice. Flowers unisexual. Calyx trifid. Ovary superior, with a single erect ovule. Fruit 2-valved. Seed enveloped in a many-parted aril; embryo very minute; albumen ruminate.

Uses.—Myristica moschata yields the well-known spices mace and nutmeg. Similar aromatic qualities pervade the order.

Typical Genus.—Myristica.

136.—*Proteaceæ.* Shrubs or small trees. Leaves hard, dry, without stipules. Calyx valvular. Stamens 4, opposite the segments of the calyx, and usually imbedded in their points. Ovary superior, simple; style simple. Fruit dehiscent or indehiscent. Seed without albumen.

Uses.—These are often handsome bushes with densely capitate flowers, and in Australia are regarded as indications of bad land; but they are of little use. The seeds of Guevina are large, almond-like, and sold as nuts in the markets of Chili.

Typical Genera.—Protea, Banksia, Grevillea.

137.—*Elæagnaceæ.* Trees or shrubs with a scurfy surface. Leaves entire, without stipules. Flowers axillary, often fragrant. Males: calyx 4-parted; stamens 3 to 8, sessile.

Female: calyx inferior, tubular, persistent. Ovary 1-celled; ovule ascending; stigma subulate. Fruit enclosed within the calyx; embryo surrounded by fleshy albumen.

USES.—The succulent fruit of Elæagnus hortensis and orientalis forms a part of an oriental dessert. That of Hippophaë rhamnoides, the Sea Buckthorn of England, may be eaten.

TYPICAL GENERA.—Elæagnus, Shepherdia.

138.—*Thymelaceæ*. Stem shrubby. Leaves without stipules. Calyx inferior, tubular, often coloured. Stamens definite, in the orifice of its tube. Ovary with one pendulous ovule. Fruit nut-like or drupaceous. Albumen none, or thin; embryo straight; radicle superior.

USES.—The bark of the species is generally caustic; that of Daphne Laureola, the Spurge Laurel, acts as a vesicant; the succulent black fruits are dangerous. Lace Bark, the liber of Lagetto lintearia, derives its name from the delicate white fibres, which are tough, and easily separated by a little violence. The same toughness of the fibre is found in many species; Daphne cannabina derives its name from being as tough as Cannabis (Hemp).

TYPICAL GENERA.—Daphne, Gnidia, Struthiola.

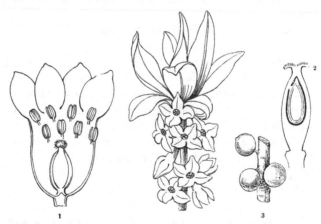

Daphne Mezereum. 1. A flower cut open. 2. A vertical section of an ovary. 3. The fruit.

139.—*Santalaceæ*. Trees, shrubs, or herbaceous plants. Leaves alternate, without stipules. Flowers small. Calyx half-coloured, valvate. Stamens 4-5, inserted in the base of the calyx. Ovary 1-celled. Ovules 1-4, attached to a central placenta. Style single. Fruit 1-seeded, indehiscent. Embryo in the axis of albumen.

USES.—The wood is sometimes fragrant; Sandal wood is obtained from several species of Santalum.

TYPICAL GENERA.—Thesium, Nyssa, Santalum.

140. — *Aristolochiaceæ.* Herbaceous plants or shrubs. Leaves alternate, often with leafy stipules. Wood without concentric zones. Flowers brown, or some dull colour, hermaphrodite. Calyx superior, valvate. Stamens epigynous. Ovary inferior, 3- or 6-celled; style simple; stigmas radiating. Fruit 3- or 6-celled, many-seeded. Seeds with a minute embryo in the base of fleshy albumen.

USES.—Many are tonic and stimulating. Aristolochia serpentaria and fragrantissima are employed as powerful aromatics; others, as A. Clematitis, indica, &c. are emmenagogues. The Asarums seem to have similar qualities, but more feeble; A. Canadense is called Wild Ginger in North America. In consequence of their stimulating properties some are employed as alexipharmics; the Guaço of the Oronoko, said to be a specific against the bite of snakes, is a species of Aristolochia.

TYPICAL GENERA.—Aristolochia, Asarum.

141.—*Empetraceæ.* Small acrid shrubs with heath-like evergreen leaves and minute flowers, which are unisexual. Sepals: hypogynous imbricated scales. Stamens equal in number to the inner sepals, and alternate with them. Ovary 3- 6- or 9-celled; ovules solitary, ascending; stigma radiating. Fruit fleshy, 3- 6- or 9-celled; the coating of the cells bony; embryo in the axis of fleshy watery albumen.

USES.—Unknown.

TYPICAL GENERA.—Empetrum, Ceratiola.

142.—*Euphorbiaceæ.* Trees, shrubs, or herbaceous plants, often abounding in acrid milk. Leaves opposite or alternate, usually with stipules. Flowers sometimes enclosed within

an involucre, monœcious or diœcious. Calyx lobed, sometimes wanting. Corolla consisting of petals, or scales, or
absent. Stamens definite or indefinite. Ovary superior,
2- or 3-celled ; ovules solitary or twin, suspended ; styles
equal in number to the cells; stigma compound or single.
Fruit generally consisting of 3 dehiscent cells, separating
with elasticity from their common axis; embryo in fleshy
albumen.

USES.—Castor-oil is obtained from the seeds of Ricinus
communis; Tiglium-oil from that of Croton Tiglium; and
a similar purgative quality seems to be general in the seeds
of the order. Cascarilla is the bark of Croton Eleutheria ;
and the same aromatic principle occurs in many species.
Many are deadly poisons, as Manchineel, Hyænanche, Sapium
aucuparium, &c. The drastic drug Euphorbium flows from
the stem of some succulent Euphorbias in North Africa.
Boxwood, so useful to wood engravers, is the timber of
Buxus sempervirens. Cassava, or Mandioc, or Tapioca, a
nutritious substance consisting of starch, is obtained from
the stem of Jatropha Manihot, a poisonous plant ; but it is
purified by washing and torrefaction.

TYPICAL GENERA.—Buxus, Andrachne, Cluytia.

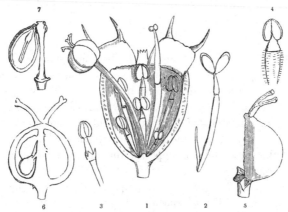

1. The involucre of a Euphorbia, containing monandrous male florets, surrounding
a long-stalked female.   2. 3. 4. Male florets of different species, with the articulation
that separates the filament from the pedicel.   5. A carpel separate.   6. A vertical
section of an ovary.   7. A vertical section of a ripe seed, showing the central column
and an embryo in the midst of albumen.

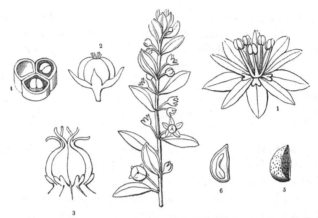

Andrachne telephioides. 1. A male flower. 2. A female flower. 3. A pistil with the scales at its base. 4. A transverse section of an ovary. 5. A ripe seed. 6. A vertical section of it.

143.—*Chloranthaceæ.* Herbaceous plants. Stems jointed. Leaves opposite with intervening stipules. Flowers hermaphrodite or unisexual. Stamens lateral; anthers 1-celled, with a fleshy connective. Ovary 1-celled. Ovule pendulous. Fruit drupaceous. Embryo minute at the apex of fleshy albumen.

Uses.—Chloranthus officinalis and others are powerful aromatics, especially the roots, which have been used with success in dangerous typhus.

Typical Genus.—Chloranthus.

144.—*Piperaceæ.* Shrubs or herbaceous plants. Leaves without stipules. Flowers usually sessile in spikes, hermaphrodite. Stamens definite or indefinite. Ovary superior, 1-celled, containing a single erect ovule ; stigma sessile, simple. Fruit somewhat fleshy, indehiscent. Seed erect, with the embryo lying in a fleshy sac or vitellus placed at that end of the seed which is opposite the hilum, on the outside of the albumen.

Uses.—The pungent aromatic peppers of the shops are obtained from different species; Piper nigrum yields black and white pepper; P. longum the long pepper. Cubebs is the pepper of P. caninum and others. P. Betel and methysticum are both intoxicating.

Typical Genera.—Piper, Peperomia.

Serronia Jaborandi. 1. A cluster of flowers magnified. 2. A ripe fruit. 3. A vertical section of the same, showing the seed and the position of the embryo.

145.—*Saururaceæ.* Herbaceous marsh or water plants. Leaves alternate, with stipules. Flowers hermaphrodite. Stamens 6, clavate, persistent. Ovaries 4, distinct, with solitary ascending ovules; or a 3- 4-celled pistil. Nuts 4, indehiscent; or a 3- 4-celled capsule. Embryo minute in a fleshy sac or vitellus, on the outside of hard mealy albumen.

Uses.—Unknown.

Typical Genera.—Saururus, Aponogeton.

146.—*Salicaceæ.* Trees or shrubs. Leaves alternate, simple, with stipules. Flowers unisexual, amentaceous. Ovary superior, 1-celled; ovules numerous, erect. Fruit coriaceous, 1-celled, 2-valved, many-seeded. Seeds comose; albumen 0.

Uses.—Various species of Salix are the Willows from whose flexible shoots wicker-work is made. S. alba is a very large fast-growing tree, and its bark abounds in tannin; S. Russelliana and purpurea yield a good febrifugal bark. The same property resides in Populus tremula, and other species of that genus; the young buds of Populus candicans and balsamifera exude a fragrant resin used in medicine; finally, the timber of Poplars is light, clean, and very useful for purposes in which hardness and strength are not required.

Typical Genera.—Populus, Salix.

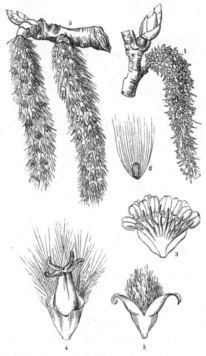

Populus. 1. Nigra. 2. Tremula. 3. A male flower. 4. A female flower. 5. A ripe capsule. 6. A seed.

147.—*Platanaceæ.* Trees or shrubs. Leaves alternate, with scarious sheathing stipules. Flowers amentaceous, in round unisexual catkins. Stamens single. Ovaries terminated by a thick style, having the stigmatic surface on one side; ovules solitary, or two, suspended. Nuts clavate. Seeds solitary; embryo in the axis of fleshy albumen.

Uses.—The large species yield a beautiful but brittle and perishable timber.

Typical Genus.—Platanus.

148.—*Urticaceæ.* Trees, shrubs or herbs, sometimes lactescent. Leaves alternate, usually covered with asperities or stinging hairs; with stipules. Flowers small, monœcious or diœcious. Calyx membranous. Stamens definite, often turned back with elasticity. Ovary superior, simple; ovule

solitary, erect or suspended ; stigma simple. Fruit, a nut.
Embryo with or without albumen ; radicle always superior.
Division 1.—*Urticeæ.* Flowers loose. Juice watery.
TYPICAL GENERA.—Urtica, Parietaria.

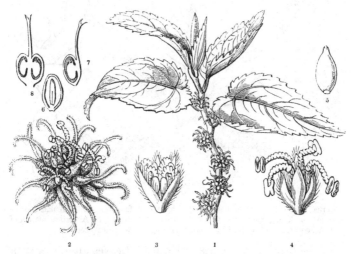

1. Branch of Procris splendens. 2. Cluster of male and female flowers. 3. A
male flower about to expand. 4. The same expanded. 5. A nut of hemp. 6. A
vertical section of it. 7. A vertical section of the ovary of Dorstenia. 8. An acci-
dental double pistil of the same.

Division 2.—*Artocarpeæ.* Flowers consolidated. Juice
milky.

TYPICAL GENERA.—Morus, Artocarpus.

USES.—The leaves of Hemp are narcotic ; and the Upas,
(Antiaris toxicaria,) certain kinds of Fig, and many Nettles,
are dangerous narcotico-acrid poisons. The deleterious principle is, however, so little developed in some that they become harmless, and are used for food, as the fruit of the
common Fig, the Mulberry, the Bread-fruit, (Artocarpus,) and
several others. Even the milky juice, which is generally
very acrid, is bland in some cases, especially that of the
Cow-tree of Equinoctial America, on which the natives feed ;
it always abounds in Caoutchouc, which is obtained in large
quantities from many kinds of Fig. The Banyan-tree of India
is Ficus indica. The toughness of fibre found in Hemp
is also common in other species, especially some nettles and

Broussonetia papyrifera. Hops, so valuable for their bitterness, consist of the bracts and ripe fruit of Humulus Lupulus.

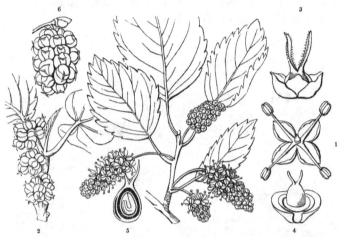

Morus alba. 1. A male flower. 2. Clusters of females. 3. A female flower separate. 4. The same with a part of the calyx cut away. 5. A vertical section of a ripe achænium. 6. A cluster of fruit consisting of baccate calyxes.

149.—*Betulaceæ.* Trees or shrubs. Leaves alternate, with stipules. Flowers unisexual, amentaceous, monœcious; males sometimes having a calyx. Stamens definite, usually distinct. Ovary superior, 2-celled; ovules pendulous. Fruit membranous, indehiscent, 1-celled. Seeds pendulous; albumen none.

Uses.—Timber trees, furnishing a light kind of wood.

Typical Genera.—Betula, Alnus.

150.—*Myricaceæ.* Leafy shrubs, with resinous glands and dots, leaves alternate. Flowers unisexual, amentaceous, achlamydeous. Stamens 6 or 8. Ovary 1-celled, surrounded by several hypogynous scales; ovule solitary, erect; stigmas 2. Fruit drupaceous, or dry and dehiscent. Seed solitary, erect; radicle superior.

Uses.—Aromatic shrubs. Sweet Gale, Myrica Gale, is used in Sweden as a substitute for hops. The berries of the M. cerifera, the Candleberry Myrtle, secrete a natural wax from their surface; its roots are astringent. Comptonia asplenifolia is used in North American medicine in diarrhœa.

Typical Genera.—Myrica, Comptonia.

151.—*Corylaceæ* or *Cupuliferæ*. Trees or shrubs. Leaves with stipules, alternate. Flowers unisexual, amentaceous. Stamens 5 to 20. Ovary crowned by the rudiments of a calyx, within a coriaceous involucre, with several cells and several ovules; ovules pendulous. Fruit a 1-celled nut, enclosed in the involucre. Seed solitary.

Uses.—The Beech, valuable for its timber, and its nuts called mast, is Fagus sylvatica. Hazel-nuts are the fruit of Corylus Avellana. Sweet Chesnuts are borne by Castanea vesca. Oaks, the most valuable of European trees, are various species of Quercus.

Typical Genera.—Fagus, Quercus.

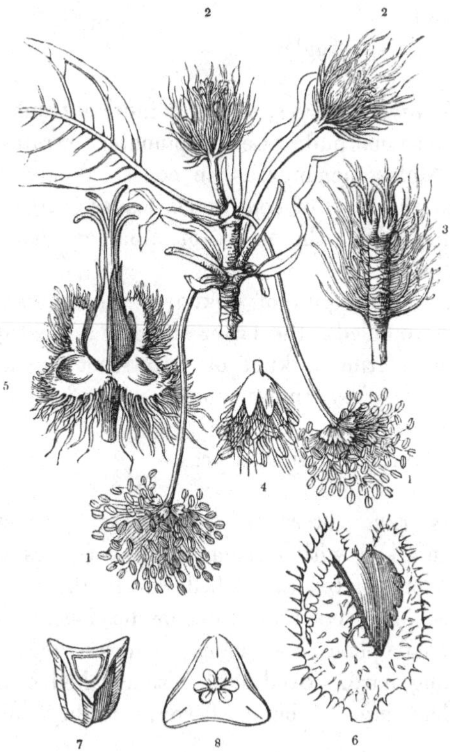

Fagus sylvatica. 1. Male catkins. 2. Female do. 3. The latter, with the scales of the involucre stripped off to show the ovaries at the apex. 4. A male flower. 5. A half-grown female, with the involucre, now consisting of consolidated scales, forced back. 6. A ripe involucre opening and exposing the nuts. 7. A transverse section of a ripe nut. 8. The same of a young ovary.

152.—*Juglandaceæ.* Trees. Leaves alternate, pinnated, without pellucid dots or stipules. Flowers unisexual, amentaceous. Calyx in the males membranous; in the females superior. Petals in the males 0; in the females occasionally present. Stamens indefinite. Ovary inferior, incompletely 2- 4-celled; ovule solitary, erect. Fruit drupaceous, 1-celled, with 4 imperfect partitions. Seed 4-lobed; radicle superior.

Uses.—Trees furnishing excellent timber; that of Juglans regia and nigra is used for gunstocks; of Carya alba for purposes of elasticity and strength: the former are Walnuts, the latter Hickory. The fruit is purgative; that of the common Walnut when young, made into a preserve with the husk, is a domestic medicine; and Juglans cathartica derives its name from its quality.

Typical Genus.—Juglans.

153.—*Cycadaceæ.* Trees, with a cylindrical trunk, increasing by a single terminal bud. Leaves pinnated, gyrate. Flowers diœcious. Males monandrous, in cones. Females either in cones, or in the form of contracted leaves. Ovules solitary, naked. Embryo in the midst of albumen, hanging by a spiral suspensor.

Uses.—A bitter gum of unknown use exudes from the trunk when wounded; the latter contains a great quantity of starch, which forms a kind of arrow-root extracted from Zamias in the West Indies, and a sort of Sago from the species of Cycas.

Typical Genera.—Zamia, Cycas.

154.—*Taxaceæ.* Trees with continuous branches. Ligneous tissue marked with circular disks. Leaves usually entire; sometimes dilated and lobed, and in those cases having forked veins. Flowers monœcious or diœcious, solitary. Filaments monadelphous. Females; ovules naked, their outer skin becoming hard. Seed hard, either naked or surrounded by a succulent cup. Albumen fleshy. Embryo dicotyledonous.

Uses.—The Yew and several others are valuable timbertrees. The leaves of Yew are fœtid and deleterious; they are said to act medicinally like Digitalis without accumulating in the system; the succulent fruit seems harmless; but the seeds are said to be dangerous.

TYPICAL GENERA.—Taxus, Dacrydium.

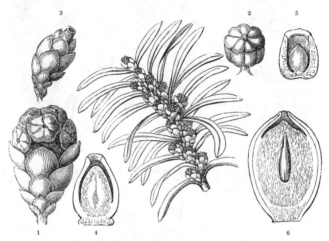

Taxus baccata loaded with male flowers. 1. A male flower. 2. An anther. 3. A female flower. 4. A vertical section of an ovule. 5. Of a ripe fruit. 6. Of a ripe seed, showing the embryo.—N.B. 4. and 6. are the same part in youth and age. 5. Is the ripe ovule, with an accessory cup.

155.—*Pinaceæ*, or *Coniferæ*. Trees or shrubs, with a branched trunk abounding in resin. Ligneous tissue marked with circular disks. Leaves entire. Flowers monœcious or diœcious. Males monandrous or monadelphous, collected in a deciduous catkin. Females in cones. Ovary a flat scale. Ovules naked. Fruit a cone. Seed with a hard integument. Embryo in oily albumen, with 2 or many opposite cotyledons.

USES.—The timber is of great value; Deal, Fir, Pine, Cedar, Larch wood are produced by various species. Turpentine, resin, pitch, and similar substances are obtained from others; the resin Sandarach exudes from Thuja articulata. Juniper-berries are the galbuli of Juniperus communis, and are diuretic. Savin, a dangerous emmenagogue, is the Juniperus Sabina. Larch bark is equal to that of Oak for tanning power.

TYPICAL GENERA.—Thuja, Abies, Cupressus.

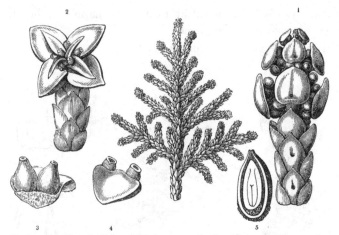

Thuja orientalis. 1. A magnified fragment of a branch bearing a cone of male
flowers. 2. A portion of a female branch. 3. 4. Scales with naked ovules. 5. A
vertical section of a ripe seed.

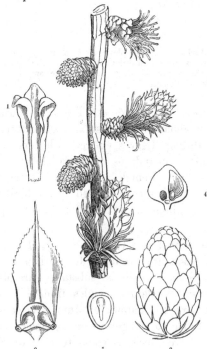

Abies Larix. 1. An anther. 2. A female scale with ovules. 3. A ripe cone.
4. A scale of the latter with a naked seed. 5. Vertical section of seed and embryo.

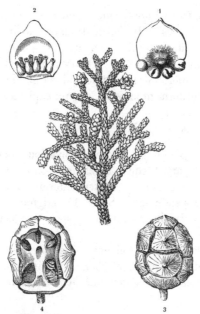

Cupressus sempervirens. 1. A scale of a male cone with pollen. 2. A scale of a female cone with naked ovules. 3. A ripe cone. 4. The same with one of the scales removed.

CLASS II. ENDOGENÆ.

This class is much smaller than the last, and much more easy to arrange systematically. For general purposes the following subdivisions may be used:

1. *Rhizanthæ.* Fungoid parasitical plants.
2. *Floridæ.* Leafy plants with the floral envelopes verticillate.
3. *Glumaceæ.* Leafy plants with the floral envelopes imbricated.

SUBCLASS I. RHIZANTHÆ.

Order 156.—*Rafflesiaceæ.* Flowers by abortion diœcious. Perianth superior, 5-parted, imbricated; the throat surrounded by calli. Column adhering to the tube of the perianth; anthers numerous, 2-celled, opening by a vertical aperture. Ovary inferior, 1-celled, with many-seeded parietal placentæ; styles conical.

Uses.—Astringents; scarcely known.

Typical Genera.—Rafflesia, Pilostyles.

157.—*Cytinaceæ.* Flowers monœcious, at the top of a stalk covered with scales. Perianth tubular, with a spreading limb. Column fleshy, thickened at the point, covered by anthers. Anthers 8, 2-celled. Ovary inferior, 1-celled, with 8 parietal placentæ. Style simple, joined to the tube of the perianth by septiform processes; stigma capitate, thick.

Uses.—Unknown.

Typical Genera.—Cytinus, Aphyteia.

158.—*Balanophoraceæ.* Flowers monœcious, in dense heads. Calyx deeply 3-parted, equal, spreading, sometimes imperfect. Stamens 1-3, epigynous. Ovary inferior, 1-2-celled, 1-2-seeded; ovule pendulous. Style 1; stigma simple, rather convex. Fruit 1-celled, containing spores collected in a bag resembling a seed.

Uses.—Cynomorium coccineum was formerly used as an astringent, under the name of Fungus Melitensis.

Typical Genera.—Balanophora, Cynomorium.

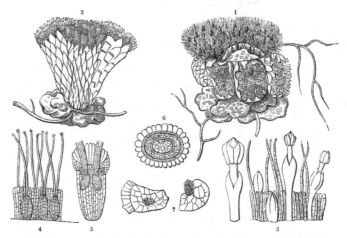

Scybalium fungiforme. 1. A male plant. 2. A female. 3. Male flowers with hairs between them. 4. Females. 5. A vertical section of a female, with the two pendulous ovules. 6. A section across a ripe fruit. 7. Seeds.

SUBCLASS II. FLORIDÆ.

159.—*Hydrocharaceæ.* Floating or water-plants. Sepals 3, herbaceous. Petals 3, coloured. Stamens definite or indefinite. Ovary 1- or many-celled; stigmas 3-6; ovules often parietal. Seeds without albumen; embryo undivided, anti-tropous.

Uses.—Unknown.

Typical Genera.—Hydrocharis, Stratiotes.

160.—*Zingiberaceæ* or *Scitamineæ*. Aromatic, tropical, herbaceous plants. Leaves with divergent veins. Calyx superior, tubular. Corolla irregular, with 6 segments in 2 whorls. Stamens 3, of which the 2 lateral are abortive. Filament not petaloid. Anther 2-celled. Stigma dilated, hollow. Fruit usually capsular, occasionally berried. Seeds with or without an aril; albumen floury; embryo enclosed within a vitellus.

Uses.—Aromatic stimulants. Ginger is the rhizoma of Zingiber officinale; Cardamoms are the fruit of Elettaria Cardamomum and others. Grains of Paradise, or Meleguetta pepper, are furnished by Amomums. Turmeric, Galangale, and Zedoary are other products of the order.

Typical Genera.—Alpinia, Hedychium.

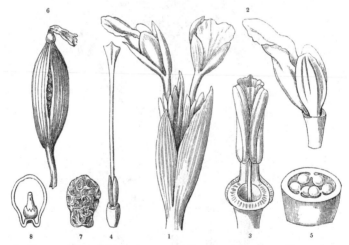

1. Flowers of Kæmpferia pandurata. 2. The inner row of the corolla seen in profile. 3. The anther, enclosing the apex of the style between its lobes. 4. The style and stigma, with two abortive stamens at the base. 5. A transverse section of the ovary. 6. Ripe fruit of Ceylon Cardamoms, Elettaria Cardamomum Zeylanicum of Pereira. 7. A seed. 8. The same cut through to show the embryo seated in vitellus.

161.—*Orchidaceæ*. Herbaceous plants, in tropical countries often growing on trees and rocks. Leaves often articulated with the stem. Sepals 3. Petals 3, of which 2 are uppermost, and 1, the lip, undermost. Stamens 3, united in a column, the 2 lateral abortive, the central perfect, or the central abortive, and the 2 lateral perfect; pollen powdery, or cohering in masses. Ovary 1-celled, with 3 parietal placentæ; style a part of the column of the stamens; stigma a viscid space in front of the column. Seeds very numerous, minute.

Uses.—The roots of Orchis mascula and others contain a large quantity of hard mucilage, and form a nutritious substance called Salep. The fragrant Vanilla is the succulent fruit of Vanilla planifolia. The corm of Bletia verecunda is bitter; the expressed juice of Epidendrum bifidum is said to be purgative.

Typical Genera.—Orchis, Epidendrum, Spiranthes, Oncidium.

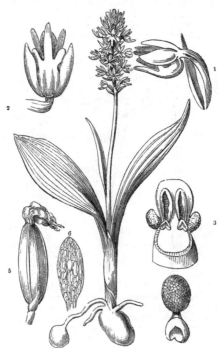

Herminium monorchis. 1. A flower seen in profile. 2. The same seen from below. 3. The column. 4. A pollen mass and gland. 5. A ripe capsule. 6. A seed with its netted purse-like skin.

162.—*Marantaceæ.* Herbaceous tropical plants destitute
of aroma. Leaves with divergent veins. Calyx superior, of
3 sepals. Corolla irregular, with the segments in 2 whorls.
Stamens 3, petaloid, of which one of the laterals and the
intermediate are barren or abortive. Filament petaloid;
anther 1-celled. Stigma cucullate, and incurved. Seeds
without aril; albumen hard; embryo naked.

Uses.—Maranta arundinacea and some others form a large
quantity of pure starch in their tubers, and this, when extracted, forms arrow-root; the leaves of both this and the
Zingiberaceous order are plaited into baskets by the Indians.

TYPICAL GENERA.—Maranta, Canna.

163.—*Musaceæ.* Leaves with divergent veins, sheathing at
the base, and forming a kind of spurious stem; often very
large. Flowers spathaceous. Perianth 6-parted, petaloid,
in 2 rows. Stamens 6, some abortive; anthers 2-celled.
Stigma usually 3-lobed. Fruit either a 3-celled capsule, or
succulent. Embryo in the axis of mealy albumen.

Uses.—The large fleshy fruits filled with starch in Musa
are the Plantains and Bananas of tropical countries, where
they furnish the inhabitants with an abundance of most nutritious food.

TYPICAL GENERA.—Musa, Strelitzia.

164.—*Amaryllidaceæ.* Generally bulbous, sometimes fibrous-
rooted, occasionally with a lofty stem. Leaves ensiform.
Calyx and corolla equally coloured, superior. Stamens 6;
anthers bursting inwardly. Stigma 3-lobed. Albumen fleshy
or corneous.

Uses. — Hæmanthus toxicarius, and many others, have
poisonous bulbs. The deleterious principle in a diffused state
renders them simply emetic, as in Narcissus, several species of
which possess this quality; or purgative, as Oporanthus luteus.
In some Alströmerias with fleshy roots a large quantity of
starch exists, which, when freed from impurities, forms a sort
of arrow-root. Agave Americana, the American *Aloe*, as it is
miscalled by gardeners, abounds, when flowering, in a sweet
sap, which, being fermented, becomes an intoxicating liquid,
called Pulque.

TYPICAL GENERA.—Amaryllis, Oporanthus, Narcissus.

N

Pancratium maritimum. 1. A flower cut open, and showing that there is a bifid tooth, forming a coronet or cup, between each stamen. 2. A transverse section of the ovary. 3. A section of the seed, showing the embryo.

165.—*Dioscoreaceæ*. Twining shrubs. Leaves alternate, netted, with a distinct petiole. Flowers minute, diœcious. Calyx and corolla superior. Stamens 6. Ovary 3-celled, with 1- or 2-seeded cells; style deeply trifid. Fruit leafy, compressed, occasionally succulent. Embryo small, near the hilum, in a large cavity of cartilaginous albumen.

Uses.—The roots of many species of Dioscorea abound in starch, and are the Yams used for food in tropical countries instead of Potatoes. Nevertheless there is present a highly deleterious principle, that, when concentrated, renders these plants dangerous. The root of Tamus communis is very acrid; and even some Yams are too nauseous to be used for food, even after careful cooking.

Typical Genera.—Tamus, Dioscorea.

166.—*Iridaceæ.* Herbaceous plants or under-shrubs. Stem often a rhizoma or cormus. Leaves usually equitant. Calyx and corolla confounded, sometimes irregular. Stamens 3, from the base of the sepals; anthers bursting externally. Stigmas 3, often petaloid. Albumen corneous, or densely fleshy.

Uses.—Crocus sativus has long orange-coloured stigmas, which, when dried, form saffron. Orris-root is the slightly stimulating aromatic rhizoma of Iris florentina and others; that of I. pseudacorus is acrid, purgative, and emetic.

Typical Genera.—Iris, Crocus, Tigridia.

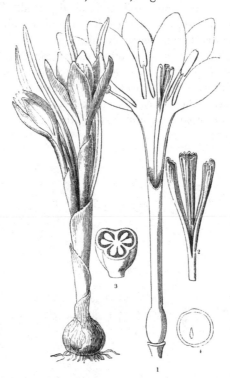

Crocus vernus. 1. A flower split open. 2. The stigmata. 3. A transverse section of the ovary. 4. A section of the seed to show the embryo.

167.—*Bromeliaceæ.* Stemless or short-stemmed plants, with rigid channelled leaves often covered with cuticular scurf. Calyx 3-parted, superior, usually herbaceous. Petals coloured. Stamens 6, or more. Stigma 3-lobed, or entire, often twisted.

Seeds numerous; embryo taper, or minute, in the base of mealy albumen.

USES.—The sub-acid fragrant fruit of Ananassa sativa is the well-known Pine-apple. The dry filamentous stems of Tillandsia usneoides are used in tropical countries for stuffing mattresses.

TYPICAL GENERA.—Bromelia, Tillandsia.

168.—*Smilaceæ*. Herbaceous plants or under-shrubs, with a tendency to climb. Stems woody. Flowers hermaphrodite or diœcious. Calyx and corolla inferior, 6-parted. Stamens 6. Ovary 3-celled; stigmas 3. Fruit a roundish berry. Albumen between fleshy and cartilaginous.

USES.—The diuretic demulcent called Sarsaparilla is the root of several species; others have a large fleshy root possessing similar properties, and called *Chinæ radix:* it appears to be nutritious. The leaves of Smilax glycyphylla are bittersweet, and are used for tea in New Holland.

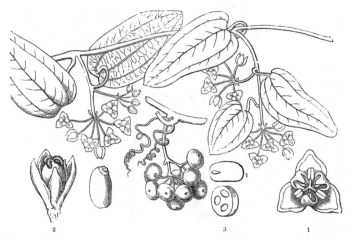

Smilax glycyphylla. 1. A male flower seen from above. 2. A female flower. 3. A transverse section of an ovary. 4. A seed. 5. A section of a seed, showing the embryo.

169.—*Liliaceæ*. Roots fibrous or fasciculate. Stem none; a bulb; or tuberous, or creeping, or arborescent. Calyx and corolla inferior, coloured, regular. Stamens 6. Anthers opening inwards. Ovary 3-celled; stigma simple, or 3-lobed. Fruit 3-celled. Embryo in the axis of fleshy albumen.

Uses.—Asparagus is the young shoots of Asparagus officinalis. Squills, so well known for their expectorant, emetic, and diuretic qualities, are the roots of Squilla maritima. What are called Alliaceous plants are found here in the form of Garlic, Onions, Chives, Leeks, and Rocambole, all species of Allium. The purgative drug Aloes is an extract from Aloe socotrina and other species. The Dragon-tree of Teneriffe is an arborescent form of the order, and yields an astringent substance called Gum Dragon.

Typical Genera.—Tulipa, Fritillaria, Hyacinthus.

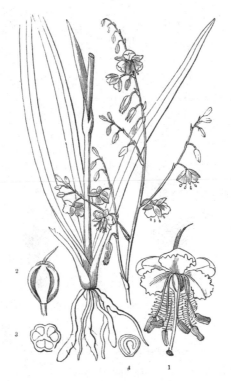

Arthropodium paniculatum. 1. A flower magnified. 2. A ripe capsule. 3. A transverse section of it. 4. A vertical section of a seed.

170.— *Melanthaceæ*. Roots fibrous, sometimes fascicled. Rhizoma sometimes a fleshy corm. Leaves sheathing at the base. Perianth inferior, in 6 pieces, or tubular. Stamens 6; anthers opening outwards. Ovary 3-celled, many-seeded; style trifid or 3-parted. Capsule divisible into 3 pieces. Albumen dense, fleshy.

Uses.—Poisonous plants. Meadow Saffron (Colchicum autumnale) is an acrid narcotic and cathartic. White Hellebore the root of Veratrum album, Cebadilla produced by Asagræa officinalis, and the roots of various kinds of Trillium and Helonias, possess similar properties.

Typical Genera.—Veratrum, Colchicum.

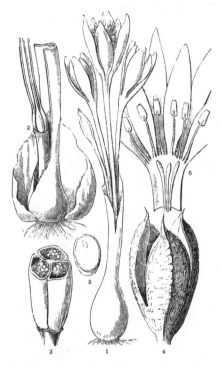

Colchicum autumnale. 1. A corm in flower. 2. The same stripped of its outer coats, and showing the ovaries after the floral envelopes are cut away. 3. A transverse section of the ovaries. 4. A ripe capsule. 5. A section of a seed. 6. The flower cut open to show the stamens and the 3-parted style.

171.—*Juncaceæ.* Herbaceous plants, with fascicled or fibrous roots. Flowers generally brown or green, hermaphrodite or unisexual. Calyx and corolla more or less glumaceous.

Stamens 6, sometimes 3. Ovary 1- or 3-celled. Stigmas generally 3. Fruit capsular, with 3 valves. Seeds neither black nor crustaceous ; albumen firm ; embryo within it.

USES.—Unimportant. Used for making mats and similar objects.

TYPICAL GENERA.—Juncus, Luzula.

172.—*Commelynaceæ*. Herbaceous plants. Leaves usually sheathing. Sepals 3, inferior, herbaceous. Petals coloured, sometimes cohering at the base. Stamens hypogynous, some deformed. Ovary 3-celled ; stigma 1. Capsule 2- or 3-celled. Seeds often twin ; embryo pulley-shaped, in a cavity remote from the hilum ; albumen fleshy.

USES.—Unknown.

TYPICAL GENERA.—Commelyna, Tradescantia.

173.—*Butomaceæ*. Aquatic plants. Leaves very cellular, often milky. Sepals 3, inferior, herbaceous. Petals 3, coloured. Stamens definite or indefinite. Ovaries 3, 6, or more. Follicles many-seeded. Seeds minute, attached to the whole of the inner surface of the fruit.

USES.—Unimportant.

TYPICAL GENERA.—Limnocharis, Butomus.

174.—*Palmaceæ*. Stem simple, rarely forked. Leaves terminal, very large, pinnate, or flabelliform, plaited in vernation. Spadix enclosed in a valved spatha. Flowers small, hermaphrodite, or polygamous. Perianth 6-parted, persistent. Stamens inserted into the base of the perianth, definite or indefinite. Ovary 3-celled, or deeply 3-lobed, with an erect ovule. Fruit baccate or drupaceous. Albumen cartilaginous or fleshy ; embryo in a cavity at a distance from the hilum.

USES.—The Cocoa-nut, whose whole structure appears useful, independently of its agreeable fruit, is the produce of Cocos nucifera ; the tough coarse fibre of this plant is manufactured into the elastic cables called Coir-rope. The Date-tree is the Phœnix dactylifera. Sago, a nutritious starchy substance, is secreted in the trunks of several species, especially of Sagus lævis and Caryota urens. The sugary nature of their sap, and its great abundance, enables the natives of Palm countries to obtain an intoxicating beverage called Palm wine from others.

The Palms of Scripture were the leaves of the Date-tree. The foliage of the order generally, being large and hard, is well suited to such purposes as thatching. The Canes, whose flexible stems when split are woven into chair-bottoms, are different species of Calamus.

Typical Genera.—Phœnix, Chamærops.

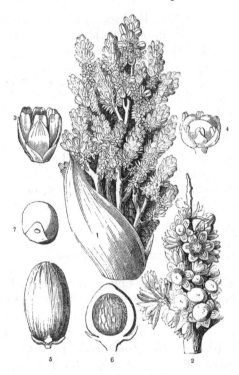

1. Inflorescence of Chamærops humilis, in its spathe. 2. A portion of the same with the fruit ripening. 3. A male flower. 4. A female flower. 5. A ripe fruit. 6. A section of another variety, showing the seed. 7. A seed with a portion of the surface cut away, to display the embryo.

175.—*Juncaginaceæ.* Herbaceous bog-plants. Leaves ensiform. Flowers inconspicuous. Sepals and petals both herbaceous, inferior, rarely absent. Stamens 6. Ovaries 3 or 6, cohering firmly; ovules 1 or 2, erect. Fruit dry; albumen wanting; embryo orthotropous, with a lateral cleft.

Uses.—Unknown.

Typical Genera.—Triglochin, Scheuchzeria.

176.—*Alismaceæ.* Floating or swamp plants. Sepals 3, herbaceous, inferior. Petals 3, petaloid. Stamens definite or indefinite. Ovaries several, 1-celled. Ovules ascending. Fruit not opening, 1- or 2-seeded. Embryo doubled upon itself.

Uses.—The leaves are acrid. The rhizoma of the Arrowhead, Sagittaria, is eatable.

Typical Genera.—Alisma, Sagittaria.

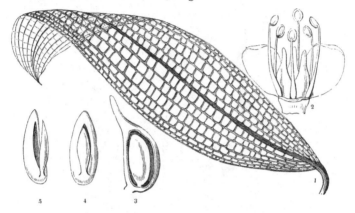

1. Leaf of Ouvirandra fenestralis. 2. A flower cut open. 3. Section of a ripe carpel of O. Bernieriana. 4, 5. Embryo in different positions: the thicker part is the cotyledon, the smaller the plumula.

177.—*Acoraceæ.* Rhizoma jointed. Leaves ensiform. Flowers hermaphrodite, surrounded with inferior scales. Spathe leaf-like. Stamens with 2-celled anthers, turned inwards. Ovaries distinct. Fruit finally juiceless. Seeds albuminous. Embryo with a cleft on one side.

Uses.—Acorus Calamus was the sweet rush with which the rooms of the higher orders were strewed before the introduction of carpets, &c. It has a fragrant rhizoma, whose aromatic qualities have rendered it useful in medicine: it is, however, chiefly employed as an ingredient in hair-powders.

Typical Genus.—Acorus.

178.—*Naiadaceæ* or *Fluviales.* Water-plants. Leaves very cellular. Flowers inconspicuous, hermaphrodite or unisexual. Perianth of 2 or 4 pieces, rarely wanting. Stamens definite. Ovaries 1 or more, superior; ovule pendulous. Fruit not opening, 1-celled, 1-seeded. Albumen none; embryo antitropous, with a lateral cleft.

Uses.—Unknown.

Typical Genera.—Potamogeton, Zannichellia.

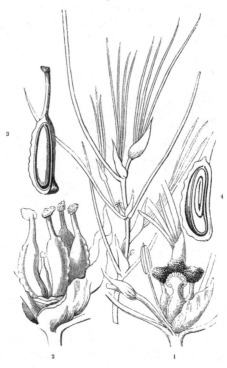

Zannichellia palustris. 1. A flower. 2. A cluster of ripe ovaries. 3. An ovary opened to exhibit the ovule. 4. A vertical section of a seed, showing the folded up embryo.

179.—*Araceæ*. Herbaceous or shrubby, stemless or arborescent plants. Leaves with parallel or branching veins; sometimes compound. Spadix generally enclosed in a spathe. Flowers unisexual. Perianth wanting. Stamens definite or indefinite, very short. Ovary 1-celled, very seldom 3-celled; ovules erect, or pendulous, or parietal. Fruit succulent. Embryo in the axis of albumen, with a cleft in one side.

Uses.—Acrid plants which are sometimes dangerous, as the Dumb cane, Dieffenbachia Seguina, which paralyses the muscles of the mouth if bitten. Nevertheless, by cooking, this acridity is so much diminished, that the leaves of Colocasia esculenta and others are used in tropical countries instead of

Cabbages. Some, too, secrete large quantities of starch, which, when separated from the acrid matter, becomes fit for food, as in Arum maculatum.

TYPICAL GENERA.—Arum, Dracontium, Caladium.

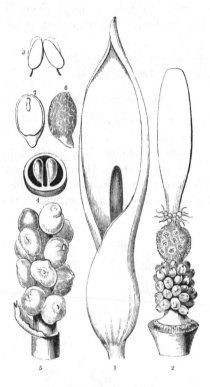

1. Spathe of Arum maculatum. 2. Its spadix loaded with flowers. 3. An anther. 4. A transverse section of an ovary. 5. A cluster of ripe fruits. 6. A seed. 7. A section of the same, showing the embryo.

180.—*Typhaceæ.* Herbaceous plants, growing in marshes or ditches. Leaves rigid, ensiform. Flowers unisexual, upon a naked spadix. Sepals 3, inferior, sometimes a bundle of hairs. Petals wanting. Stamens 3 or 6; anthers wedge-shaped. Ovary single, 1-celled; ovule pendulous; stigmas 1 or 2, linear. Fruit not opening. Embryo in the centre of albumen, with a cleft in one side.

USES.—Unknown.

TYPICAL GENERA.—Typha, Sparganium.

181.—*Pistiaceæ.* Floating plants, with very cellular, lenti-cular, or lobed stems and leaves. Flowers from the margin of the stems, inconspicuous, naked. Stamens definite. Ovary superior, 1-celled, with erect ovules. Fruit membranous or capsular, 1- or more seeded. Embryo either in the axis of fleshy albumen, and having a lateral cleft, or at the apex of the nucleus.

USES.—Acrid plants of no importance.

TYPICAL GENERA.—Lemna, Pistia.

SUBCLASS III. GLUMACEÆ.

182.—*Cyperaceæ.* Leaves with their sheaths entire. Stem solid. Flowers consisting of imbricated solitary bracts. Pe-rianth none. Stamens definite, 1, 2, 3, 4, 5, 6, 7, 10, 12; anthers fixed by their base. Ovary often surrounded by bristles; ovule erect; style single, trifid or bifid. Nut crus-taceous or bony. Embryo lenticular, within the base of the albumen.

Scirpus lacustris. 1. A flower surrounded with hypogynous bristles. 2. A seed. 3. A section of it, showing the lenticular embryo.

Uses.—Of no other importance than as covering many situations with a coarse herbage containing but little nutritive matter. A quantity of starch secreted in the tubers of some species renders them eatable, as Cyperus esculentus and others. The Papyrus of the ancients was made from the stems of the Papyrus antiquorum. A few species are slightly aromatic.

TYPICAL GENERA.—Scirpus, Schœnus, Carex.

183.—*Graminaceæ.* Stems cylindrical, usually fistular. Leaves alternate, with a split sheath. Flowers in little locustæ, consisting of imbricated bracts, with distinct glumes or paleæ, or both. Hypogynous scales 2 or 3, sometimes wanting. Stamens hypogynous, 1, 2, 3, 4, 6, or more ; anthers versatile. Ovary simple ; styles 2, very rarely 1 or 3 ; stigmas feathery. Pericarp membranous. Albumen farinaceous ; embryo on one side of the albumen, lenticular.

Uses.—The most important of all orders, because the floury albumen of certain species furnishes man with bread, and the nutritious herbage of others is the sustenance of herbivorous animals. To the class of Corn belong Wheat, Barley, Rye, Oats, Maize, Rice, and many other species cultivated in warmer countries ; to that of fodder, Crested Dogstail, various kinds of Fescue, Foxtail, Rye Grass, and a number of others cultivated by farmers. Sugar is obtained from the juice of the Saccharum officinarum, whose stem is solid, contrary to the custom of the order. Bamboos, whose hard stems are so valuable in hot countries, are arborescent grasses growing 60 to 100 feet high and more. A fragrant principle is found in Anthoxanthum odoratum and others, especially Andropogon Schœnanthus, called Lemon-grass in the gardens, which is used as a stomachic in India ; where also an oil, valued as an external application in rheumatism, is obtained from the Andropogon Calamus aromaticus, believed to have been the ancient drug of that name. The diseased grain of Rye is Ergot, valuable for its powerful action upon the uterus. Finally, a narcotic quality has been remarked in a few species, especially Lolium temulentum.

TYPICAL GENERA.—Agrostis, Bromus, Aira, Lolium.

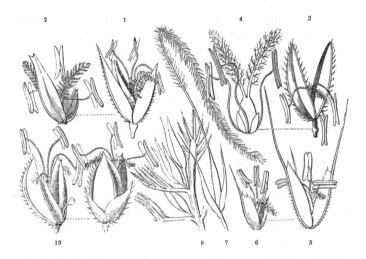

1. Locusta of Agrostis alba. 2. Paleæ and stamens, &c. of the same. 3. Paleæ of Leersia oryzoides. 4. Pistil, stamens, and hypogynous scales of the same. 5. Locusta of Polypogon monspeliensis. 6. Paleæ, &c. of the same. 7. Locusta of Stipa pennata. 8. Rachis, bracteæ, and florets of Cynosurus cristatus. 9. Locusta of Cynodon Dactylon. 10. Paleæ, and abortive floret of the same.

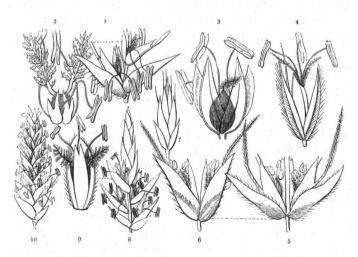

1. Locusta of Corynephorus canescens. 2. Paleæ, &c. of the same. 3. Locusta of Phalaris aquatica. 4. Locusta of Alopecurus pratensis. 5. Locusta of Aira caryophyllea. 6. Floret of the same. 7. Locusta of Festuca duriuscula. 8. Locusta of Glyceria fluitans. 9. Floret of the same. 10. Locusta of Eragrostis poæformis.

CLASS III. ACROGENS.

Substance of the plant composed of cellular tissue chiefly ; spiral vessels or ducts only present in the highest orders. Stem either increasing by an extension of its point, or by a developement in all directions from one common point ; not increasing in thickness when once formed. Sexual organs absent. Reproduction taking place by spores, or by a mere dissolution of the utricles of tissue.

184. — *Lycopodiaceæ.* Plants, with creeping stems, the axis abounding in annular ducts. Organs of reproduction axillary sessile thecæ, containing either minute powdery matter, or sporules, marked at the apex with three minute ridges.

Uses.—Some are powerful emetics and cathartics, especially L. Selago and rubrum.

Typical Genera.—Lycopodium, Bernhardia.

185.—*Filicales* or *Filices.* Leafy plants producing a rhizoma. Leaves usually coiled up in vernation, with dichotomous veins of equal thickness. Thecæ or sporangia arising from the veins upon the leaves, pedicellate with an elastic ring, or sessile and destitute of a ring.

Division 1.—*Polypodiaceæ.* Thecæ with a vertical, usually incomplete ring ; bursting irregularly and transversely.

Division 2.—*Gleicheniaceæ.* Thecæ with a transverse, occasionally oblique ring, nearly sessile, and bursting lengthwise internally.

Division 3.—*Osmundaceæ.* Thecæ with an operculiform ring, or without any ; reticulated, striated with rays at the apex ; bursting lengthwise, and usually externally.

Division 4.—*Danæaceæ.* Thecæ sessile, without any ring, concrete into multilocular sub-immersed masses, opening at the apex.

Division 5.—*Ophioglossaceæ.* Thecæ single, roundish, coriaceous, opaque, without ring or cellular reticulation, half 2-valved. Vernation straight.

Uses.—The rhizomata of some are astringent ; that of Nephrodium Filix mas has been used as an anthelmintic. In some countries the pith of the stem is used as food by the natives, especially in the islands of the South Seas.

Typical Genera.—Polypodium, Pteris, Adiantum.

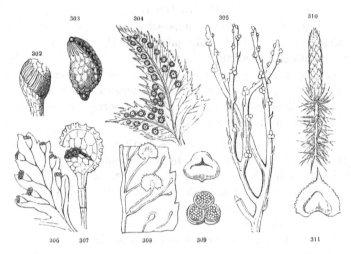

302. Theca of Trichopteris excelsa (*Gleicheniaceæ*). 303. The same of Schizæa pectinata (*Osmundaceæ*). 304. Part of the frond of Aspidium Lonchitis (*Polypodiaceæ*). 305. Bernhardia dichotoma (*Lycopodiaceæ*). 306. Part of frond of Davallia pyxidata (*Polypodiaceæ*). 307. Theca of the same. 308. Part of frond of Aspidium exaltatum. 309. Theca of Bernhardia dichotoma (*Lycopodiaceæ*). 310. Lycopodium annotinum. 311. A scale and theca of the same.

186.—*Equisetaceæ.* Leafless branched plants with a striated fistular stem; the articulations separable, and surrounded by a toothed sheath. Spiral vessels very few. Inflorescence consisting of peltate scales. Reproductive bodies in the inside of the lobes of the scales. Four clavate bodies, wrapped round a naked spore.

Uses.—The hard, flinty skin renders them fit for polishing purposes, for which some are used under the name of Dutch rushes.

Typical Genus.—Equisetum.

187.—*Characeæ.* An axis, consisting of parallel tubes. Organs of reproduction : succulent globules, containing filaments and fluid, and axillary nucules, formed of short tubes, twisted spirally.

Uses.—Unknown.

Typical Genera.—Chara, Nitella.

188.—*Bryaceæ*, or *Musci.* Cellular plants, having a distinct axis, covered with minute leaves. Reproductive organs of two kinds : viz. axillary, cylindrical stalked sacs, containing a multitude of particles emitted upon the application of water ; and thecæ or hollow urn-like cases, covered by a calyptra, closed by a lid, within which are rows of processes, called the peristome ; the centre of the theca occupied by a columella. Sporules, when germinating, protruding confervoid filaments, which afterwards ramify, and form an axis.

USES.—Unknown.

TYPICAL GENERA.—Hypnum, Bryum.

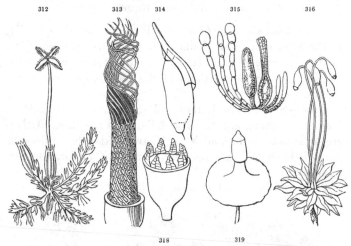

312. Jungermannia bicuspidata (*Jungermanniaceæ*). 313. Peristome of Tortula ruralis (*Muscaceæ*). 314. Theca of Ceratodon purpureus (*Musc.*) 315. Supposed representatives of sexual organs in Meesia longiseta (*Musc.*) 316. Bryum roseum (*Musc.*) 318. Peristome of Octoblepharum albidum (*Musc.*) 319. Apophysis and theca of Splachnum luteum.

189.—*Andræaceæ.* Branching moss-like plants, with imbricated leaves. Thecæ with a calyptra, splitting longitudinally into four valves. Peristome 0. Spores attached to a central columella.

USES.—Unknown.

TYPICAL GENUS.—Andræa.

190.—*Jungermanniaceæ.* Creeping moss-like plants, either with imbricated leaves, or with the leaves and axis all fused into one. Thecæ without an operculum, 4-parted, or 2-4-valved. Spores mixed with elaters.

USES.—Unknown.

TYPICAL GENUS.—Jungermannia.

191.—*Marchantiaceæ* or *Hepaticæ*. Plants composed entirely of cellular tissue, emitting roots from their under side, and consisting of an axis, bordered by a membranous expansion, which sometimes forms a broad lobed thallus. Reproductive organs consisting of a peltate stalked receptacle, bearing thecæ on its under surface; or of sessile naked thecæ, immersed, or superficial.

Uses.—Unknown.

Typical Genera.—Marchantia, Riccia.

192.—*Lichenaceæ* or *Lichenes*. Perennial plants spreading in the form of a lobed thallus. Reproductive matter of two kinds: 1, sporules lying in membranous tubes, immersed in shields; 2, separated cellules of the medullary layer of the thallus.

Uses.—Several are bitter, and some have been used as tonics; as Variolaria faginea, and Parmelia parietina. Others are nutritious, as Iceland Moss, Cetraria islandica. Roccella tinctoria is Orchal, and Lecanora Perellus, Cudbear, used extensively by dyers.

Typical Genera.—Parmelia, Lecidea, Peltidea.

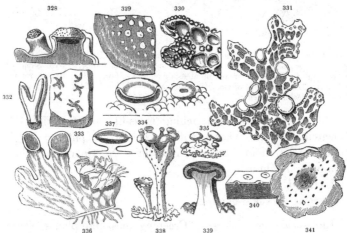

328. Shields of Variolaria amara.   329. A portion of the thallus of the same plant.
330. A piece of the thallus of Sticta pulmonacea with lacunæ and soredia.   331.
Thallus of the same, bearing shields.   332. Shield of Opegrapha scripta.   333. Thallus of the same.   334. Shields young and old of Lecanora Perellus.   335. Shields of Bæomyces rufus.   336. Part of thallus of Peltidea canina.   337. Section of a shield of Sticta pulmonacea.   338. Podetia of Cenomyce coccinea.   339. Section of shield of Bæomyces rufus.   340. Shields of Endocarpon miniatum.   341. Thallus of the same.   Chiefly from Greville's Flora Edinensis.

193.—*Algaceæ* or *Algæ.* Leafless plants, with no distinct axis ; growing in water, consisting either of simple vesicles, or of articulated filaments, or of lobed fronds. Reproductive matter either wanting or in the joints of the filaments, or in thecæ of various forms. Spores in germination elongating in two opposite directions.

USES.—A nutritious gelatinous matter is obtained from certain Gracilarias, and Chondrus crispus, sometimes called Irish Moss. Gigartina Helminthochorton has been employed as an anthelmintic. They are generally collected under the name of Wrack for burning for Kelp, formerly the source of Carbonate of Soda. The substance sold in the shops under the name of Laver is the Porphyra laciniata, and vulgaris, and the Ulva latissima.

TYPICAL GENERA.—Fucus, Conferva.

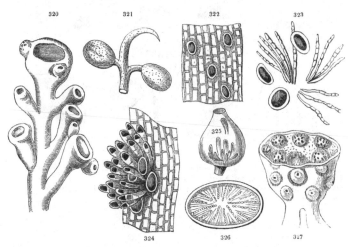

Reproductive organs of 320. Lawrencia pinnatifida. 321. Vaucheria geminata. 322. Dictyosiphon fœniculaceus. 323. Bonnemaisonia asparagoides. 324. Asperococcus echinatus. 325. Odonthalia dentata. 326. Lichina confinis. 327. Fucus vesiculosus. All from Greville's Algæ Britannicæ.

194.—*Fungaceæ* or *Fungi.* Plants consisting of cellules, among which filaments are occasionally intermixed, increasing in size by addition to their inside; their outside undergoing no change after its first formation, frequently ephemeral. Spores lying either loose among the tissue, or enclosed in sporidia.

Uses.—Agaricus campestris, the common Mushroom, and some other species of the same genus, Tuber cibarium, the Truffle, and many others, are eatable and nutritious. Others are dangerous poisons. Amadou is made from Boletus igniarius. Great numbers are mischievous parasites, infesting both live and dead organized matter, and even attacking living insects. Vast damage is committed by them under the name of Mildew, Rust, Brand, Smut, and Dry-rot.

Typical Genera.—Agaricus, Geastrum, Mucor, Hypoxylon.

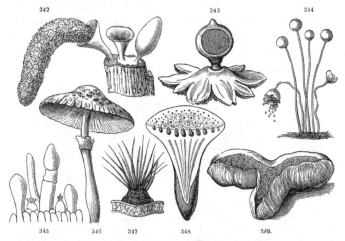

342. Arcyria flava.   343. Geastrum multifidum.   344. Mucor caninus.   345. Basidia and cystidia of an Agaric.   346. Agaricus cepæstipes.   347. Vermicularia trichella.   348. Vertical section of Hypoxylon punctatum.   349. Angioridium sinuosum.   From Greville's Cryptogamic Flora.

V. THE ALLIANCES OF PLANTS.

THE following pages explain the author's own views of arrangement in 1836, and serve as a key to the Natural System of Botany (Edition 2, London 1836). Although his opinion is much modified by subsequent consideration, yet he knows from experience that these short characters are of considerable value to students.

CLASSES.

The whole vegetable kingdom is divisible into five principal classes, which may be characterised as follows :

Propagated by sexes
- having spiral vessels
  - Exogens with their seeds in an ovary . I. EXOGENÆ.
  - Exogens with naked seeds . . II. GYMNOSPERMÆ.
  - Endogens . . III. ENDOGENÆ.
- without spiral vessels, or with scarcely any . . . . . IV. RHIZANTHÆ.

Propagated without sexes . . . . V. ACROGENÆ.

They are further known by a separate consideration of the nature of all their principal organs, thus :

	Wood.	Veins of Leaves.	Floral Envelopes.	Sexes.	Embryo.
I. EXOGENÆ	Exogens	Netted	Quinary	Perfect	Dicotyledonous
II. GYMNOSPERMÆ	Exogens	Parallel or forked	None	Imperfect	Dicotyledonous
III. ENDOGENÆ	Endogens	Parallel	Ternary	Perfect	Monocotyledonous
IV. RHIZANTHÆ	None	None	Variable	Imperfect	Acotyledonous
V. ACROGENÆ	Acrogens	Forked, or 0	None	None	Acotyledonous

The five classes form a circle, the centre of whose circumference is occupied by Exogens and Endogens, the common point by Acrogens, and the intermediate spaces by Gymnospermæ and Rhizanthæ, which are transition classes. This may be expressed thus :

Exogens, Endogens,
Gymnospermæ, Rhizanthæ,
Acrogens.

This proposition is to be demonstrated in the course of the following explanation of the characters and affinities of the various Classes, Subclasses, Groups, Alliances, and Natural Orders, of which the vegetable kingdom consists.

CLASS I. EXOGENÆ.

The Subclasses are

COMPLETE PLANTS; with both their calyx and corolla perfect; or at least with the calyx highly developed, if the petals are absent: these divide into

1. POLYPETALÆ, with the petals distinct.
3. MONOPETALÆ, with the petals united into a tube.

2. INCOMPLETE PLANTS; in which there is no corolla; their calyx is generally either but little developed or altogether absent.

No division of Exogens has been discovered more in accordance with natural affinities, than that which depends upon the different degree of developement of the flower; it is true, indeed, that its characters are not always constant, and that practical difficulties arise from the circumstance of some genera belonging to polypetalous orders having no petals, while a portion of some monopetalous orders are actually polypetalous, and so on. Nevertheless the arrangement founded upon the distinctions above recorded appears to be natural, if the latter are rightly considered.

As understood by me, all those orders in which the floral envelopes are herbaceous, and imperfectly developed, belong to Incompletæ, whether there are two rows or not, as Menispermaceæ: nor ought others, as Euphorbiaceæ, to be removed from Polypetalæ; because, although the mass of such orders is polypetalous, certain European genera, with which we are best acquainted, have no petals. With regard to those polypetalous orders, in some genera of which the petals cohere by their edges, so as to resemble a monopetalous corolla, the only means of recognising them is by observing that their petals are scarcely joined at the base; there is this, however, which assists in removing the difficulty: in true monopetalous orders the style is scarcely ever divided, except just at the point, and their fruit is therefore, in all cases, syncarpous; while, in those polypetalous genera, which take on a monopetalous appearance, the fruit is in reality apocarpous, as is the case with Anonaceæ, Crassulaceæ, Leguminosæ, Meliaceæ, Rutaceæ, &c. The two latter, although syncarpous when young, yet become truly apocarpous as their fruit ripens.

SUBCLASS I. POLYPETALÆ.

These comprehend the following groups:

1. *Albuminosæ.* Embryo very considerably shorter and smaller than the albumen.

2. *Epigynosæ.* Ovary inferior, usually having an epigynous disk.

3. *Parietosæ.* Placentæ parietal.

4. *Calycosæ.* Calyx incompletely whorled; two of the sepals being exterior.

5. *Syncarposæ.* None of the characters of the other groups, and with the carpels compactly united.

6. *Gynobaseosæ.* Carpels not exceeding five, diverging at the base, arranged in a single row around an elevated axis, or gynobase. Stamens usually separate from the calyx.

7. *Apocarposæ.* None of the characters of the other groups, but with the carpels distinct; or separable by their faces; or solitary.

Albumen none; or in small quantity.

N.B.—In the succeeding pages the first column contains a brief character of the Natural Order; the second the name of the Order; the third its sensible properties, with some officinal example in italics within brackets, when any is to be found. When the third column is blank, nothing is known of the sensible property.

GROUP I. ALBUMINOSÆ.

Alliance 1.—*Ranales.* Herbaceous plants, either apocarpous, or with parietal placentæ.

Floral envelopes in threes or fives. Sap transparent.	1. Ranunculaceæ	. Acrid, poisonous (*Black Hellebore, Aconite*).
	§ Podophylleæ	. Cathartic.
Floral envelopes in twos or fours. Sap usually milky	2. Papaveraceæ	. Narcotic (*Poppy*).
	§ Fumarieæ	. Diaphoretic and aperient.
Embryo enclosed in a vitellus. Floaters.	3. Nymphæaceæ § Hydropeltideæ.	. Slightly astringent.
Ovaries concealed in a fleshy receptacle. Floaters.	4. Nelumbiaceæ	. Wholesome.
Stamens perigynous	5. Cephalotaceæ.	

Alliance 2.—*Anonales.* Apocarpous woody plants.

Flowers unisexual, three-lobed. Stamens columnar.	6. Myristicaceæ	. Acrid, aromatic (*Nutmeg*).
Leaves with stipules, without dots.	7. Magnoliaceæ	. Bitter, tonic.
Leaves with stipules and transparent dots.	8. Winteraceæ	. Aromatic, stimulant (*Winter's Bark*).
Flowers hermaphrodite, three-parted. Stamens usually distinct.	9. Anonaceæ	. Aromatic (*Piper æthiopicum*).
Leaves without stipules. Flowers pentapetalous.	10. Dilleniaceæ	. Astringent.

Alliance 3.—*Umbellales.* Flowers epigynous, arranged in umbels. Stems usually hollow.

Carpels two	11. Umbelliferæ or Apiaceæ.	Herb poisonous (*Hemlock*); sometimes stimulant and eatable (*Parsley, Parsnip*); fruit aromatic (*Anise*).
Carpels more than two	12. Araliaceæ	. Slightly stimulant (*Ginseng*).

Alliance 4.—*Grossales.* Flowers epigynous, arranged in racemes. Stems solid.

Placentæ parietal . .	13. Grossulaceæ .	Tonic, or harmless (*Black Currants*).
Placentæ central, many-seeded.	14. Escalloniaceæ.	
Placentæ central, few-seeded	15. Bruniaceæ.	

Alliance 5.—*Berberales.* Apocarpous, with the valves of the anthers curved backwards.

16. Berberaceæ .	Acid, astringent (*Berberry*). Dye
§ Nandineæ.	yellow.

Alliance 6.—*Pittosporales.* Syncarpous, with hypogynous stamens, and the placentæ in the centre of the fruit.

Fruit two-celled. Seeds few.	17. Vitaceæ . .	Acidity and sugar (*Vine*).
Fruit with more than two cells. Seeds numerous. Stamens all perfect.	18. Pittosporaceæ.	
Petals split. Flowers unsymmetrical.	19. Olacaceæ.	
Fruit with more than two cells. Seeds numerous. Stamens half sterile.	20. Francoaceæ.	
Stigma leafy, peltate .	21. Sarraceniaceæ.	

A group in appearance natural, and agreeing with its technical character in all respects, with the exception of Nelumbium, which has no albumen ; and the genus Berberis, in which the embryo is much larger in proportion to the albumen than in any other instance.

Some connecting links are obviously wanting in this group ; and, until it is well considered, it will appear less natural than it really is, especially if we compare such plants as the Vine with the Crowfoot, or either with Sarracenia. Nevertheless, it is to be observed, that it very nearly agrees with De Candolle's Thalamiflorous subclass, and that the mutual affinities of the alliances may be demonstrated. Take Anonales and Ranales for the centre of the circumference of a circle composed of the six foregoing alliances :—

Then Anonales pass into Pittosporales through Cheiranthera ;

Pittosporales	— Grossales	—	Ribes ;
Grossales	— Berberales	—	Berberis ;
Berberales	— Umbellales	—	Nandineæ ;
Umbellales	— Ranales	—	Thalictrum ;
Ranales	— Anonales	—	Magnolia ;

and the relative position of the alliances will be thus :—

Anonales . . . . Ranales
Pittosporales   Umbellales
Grossales . . . . Berberales.

There is no difficulty in pointing out the various gradations that connect the genera belonging to the orders comprehended in the Albuminous group. The most paradoxical part of the combination is the union of baccate-fruited with dry-fruited plants: but even Vitaceæ pass into Umbelliferæ through Leea; and the petals of the genus Vitis itself are inflected at the points, in the way of Umbelliferæ.

GROUP II. EPIGYNOSÆ.

Alliance 1.—*Onagrales.* Æstivation of corolla not valvate. Placentæ central. Every part of the flower some regular multiple of two.

	22. Onagraceæ	. None. Harmless.
	§ Circæeæ.	
	§ Halorageæ	. None.

Alliance 2.—*Myrtales.* Æstivation of corolla not valvate. Placentæ central. Number of parts of the flower uncertain.

Carpels single. Petals broad	23. Combretaceæ	. Astringent (*Myrobalan*).
Carpels single. Petals very narrow.	24. Alangiaceæ	. Hydragogues.
Stipules between the leaves	25. Rhizophoraceæ	. Astringent (*Mangrove*).
Stamens bent downwards. Leaves one-ribbed.	26. Memecylaceæ.	
Stamens bent downwards. Leaves three- or more-ribbed.	27. Melastomaceæ	. Slightly astringent.
Leaves dotted, with an intramarginal vein.	28. Myrtaceæ	. Aromatic stimulant (*Cajeputi, Cloves*); bark astringent.
Leaves alternate. Flowers irregular.	29. Lecythidaceæ	. Fruit eatable (*Brazil nut*).
Leaves not dotted. Stamens straight.	30. Philadelphaceæ.	

Alliance 3.—*Cornales.* Æstivation of corolla valvate.

Leaves with stipules	. 31. Hamamelaceæ.	
Leaves without stipules	. 32. Cornaceæ .	. Tonic.
Parasitical plants, bearing their stamens on their petals.	33. Loranthaceæ	. Astringent.

Alliance 4.—*Cucurbitales.*   Placentæ parietal.

Flowers unisexual . .	34. Cucurbitaceæ	. Purgative (*Colocynth*); or eatable (*Melon, Gourd*).
Flowers with a ring of abortive stamens.	35. Loasaceæ .	. Stinging.
Petals extremely numerous.	36. Cactaceæ .	. Subacid; wholesome.
Sepals and petals alike. Glands between the stamens.	37. Homaliaceæ.	

Alliance 5.—*Ficoidales.*   Petals extremely narrow and numerous.

        38. Mesembryaceæ .   Wholesome.

Alliance 6.—*Begoniales.*   Flowers unisexual.   Placentæ central.

        39. Begoniaceæ .   Slightly astringent.

These plants seem to be all connected by a general natural relationship; and yet it is extremely difficult to fix the limits of their alliances. They appear to be connected with the Syncarpous group through Melastoma and Lythraceæ, and with the Albuminous group by the genus Eupomatia, and even by Cactaceæ, which evidently touch upon Grossulaceæ. They also pass into Monopetalæ by Melastomaceæ, which join them with Gentianaceæ. I entertain no doubt about this being nearly the true position of Begoniaceæ.

GROUP III. PARIETOSÆ.

Alliance 1.—*Cruciales.*   Embryo curved.   Albumen absent.

Stamens tetradynamous .	40. Cruciferæ or Brassicaceæ.	Pungent, stimulant (*Mustard*).
Stamens indefinite . .	41. Capparidaceæ	. Stimulant, sometimes poisonous.
Fruit composed of three carpels.	42. Resedaceæ	. None.

Alliance 2.—*Violales.*   Stamens few, with no coronet to the flower.

Leaves with stipules .	43. Violaceæ .	. Roots emetic (*White Ipecacuanha*).
Leaves dotted . .	44. Samydaceæ.	
Fruit siliquose . .	45. Moringaceæ	. Pungent, aromatic.
Leaves circinate when young	46. Droseraceæ	. Subacrid.
Calyx ribbed . . .	47. Frankeniaceæ.	

Alliance 3.—*Passionales.* Flowers with a ring or coronet of sterile stamens. Petioles generally glandular.

Leaves with stipules .	48. Passifloraceæ .	Subacid.
Flowers unisexual . .	49. Papayaceæ .	Vermifugal.
Placentæ spread over all the lining of the fruit.	50. Flacourtiaceæ .	Suspicious.
Stipules absent. Ovary stalked.	52. Malesherbiaceæ.	
Stipules absent. Ovary sessile (Coronet 0).	53. Turneraceæ.	

Alliance 4.—*Bixales.* Polyandrous. Leaves dotted.

54. Bixaceæ . .	Purgative and stomachic (*Arnotto*).

This is connected with the Epigynous group by Passiflora, and with the Calycose by Turnera, which passes into Cistaceæ. Otherwise its external relationships are not well marked. The orders themselves are intimately related.

GROUP IV. CALYCOSÆ.

Alliance 1.—*Guttales.* Polyandrous. Albumen absent. Petals equal in number to the sepals.

Leaves simple. Seeds few.	55. Guttiferæ or Clusiaceæ.	Fruit sometimes eatable (*Mangostan*); purgative, acrid (*Gamboge*).
Leaves compound. Seeds few.	56. Rhizobolaceæ .	Seeds eatable (*Sapocaya nuts*).
Leaves alternate. Flowers unsymmetrical. Seeds numerous.	57. Marcgraaviaceæ.	
Styles several. Seeds numerous.	58. Hypericaceæ .	Slightly purgative and febrifugal.

Alliance 2.—*Theales.* Polyandrous. Albumen absent. Petals unequal to the sepals in number.

59. Ternströmiaceæ .	Subnarcotic and astringent (*Tea*).

Alliance 3.—*Acerales.* Stamens definite. Flowers unsymmetrical.

Petals without appendages. Fruit indehiscent, winged, consisting of two carpels.	60. Aceraceæ .	Saccharine (*Sugar maple*).

Petals having scales in front. Fruit indehiscent, consisting of three carpels. A disk.	61. Sapindaceæ	. Leaves and branches poisonous, fruit eatable (*Litchi*).
Petals without appendages. Fruit dehiscent.	62. Æsculaceæ	. Bark astringent, febrifugal (*Horsechestnut*).
Flowers papilionaceous .	63. Polygalaceæ	. Bitter, emetic, &c.
Flowers spurred . .	64. Vochyaceæ	. Astringent (*Ratanhia root*).

Alliance 4.—*Cistales.*   Flowers regular.   Albumen present.

Stamens equal to the number of sepals.	65. Elatinaceæ.	
Decandrous, without stipules.	66. Linaceæ .	. Mucilaginous, tough (*Flax*).
Decandrous, with stipules	67. Hugoniaceæ.	
Polyandrous, with an involucre.	68. Chlenaceæ.	
Polyandrous. Style simple. Radicle remote from the hilum.	69. Cistaceæ .	. Balsamic (*Labdanum*).
Polyandrous. Styles many. Seeds hairy.	70. Reaumuriaceæ	. Saline.

The characters of this group require careful consideration. Many gynobaseous plants have a calyx imbricated in a similar way, but they are removed by their gynobasic structure. The imbricated character of the calyx depends upon this ; that the whorl of floral leaves is broken, so that about two of the sepals are out of the place of the others, and are, consequently, altogether external.

The Calycose passes into the Parietose group by Turnera, and into the Syncarpous by Hugoniaceæ.

GROUP V. SYNCARPOSÆ.

Alliance 1.—*Malvales.*   Æstivation of calyx valvate ; carpels four or more.

Stamens monadelphous. Anthers two-celled.	71. Sterculiaceæ	. Mucilaginous.
Stamens monadelphous. Anthers one-celled.	72. Malvaceæ .	. Mucilaginous (*Marsh mallow*).
Anthers bursting by pores. Petals lacerated.	73. Elæocarpaceæ.	
Stamens monadelphous. Calyx irregular and enlarged in the fruit.	74. Dipteraceæ	. Resinous (*Camphor*).
Stamens distinct, separate from calyx.	75. Tiliaceæ .	. Mucilaginous.
Stamens distinct, growing on a tubular calyx.	76. Lythraceæ	. Astringent, acrid.

Alliance 2.—*Meliales.* Æstivation of calyx imbricated; carpels four or more.

Stamens combined into a tube. Seeds wingless.	77. Meliaceæ	.	Tonic and stimulant (*Canella*).
Stamens somewhat monadelphous. Seeds winged.	78. Cedrelaceæ	.	Ditto.
Stamens monadelphous, with a dilated connective.	79. Humiriaceæ	.	Balsamic.
Leaves dotted. Fruit succulent.	80. Aurantiaceæ	.	Subacid, fragrant (*Orange*).
Stamens growing to the calyx. Disk very large.	81. Spondiaceæ	.	Harmless.

Alliance 3.—*Rhamnales.* Æstivation of calyx valvate; carpels fewer than four.

Stamens opposite the petals	82. Rhamnaceæ	.	Dye (*French berries*); purgative (*Buckthorn*).
Stamens alternate with the petals.	83. Chailletiaceæ	.	Poisonous.
Anthers opening by pores. Seeds carunculate (537).	84. Tremandraceæ.		
Somewhat polyandrous. Leaves succulent.	85. Nitrariaceæ	.	Saline.
Secreting balsam . .	86. Burseraceæ	.	Balsamic (*Balm of Gilead*).

Alliance 4.—*Euphorbiales.* Æstivation of calyx imbricated; carpels fewer than four.

Flowers unisexual. Fruit tricoccous.	87. Euphorbiaceæ	.	Stimulant, purgative, poisonous (*Castor oil, Cascarilla,* &c.)
Flowers hermaphrodite. Petals united.	88. Stackhousiaceæ.		
Seeds indefinite. Petals united.	89. Fouquieraceæ.		
Flowers hermaphrodite. Petals distinct.	90. Celastraceæ	.	Fruit sometimes eatable.
Leaves compound, with common and partial stipules.	91. Staphyleaceæ.		
Petals unguiculate. Fruit winged.	92. Malpighiaceæ	.	Fruit sometimes eatable.

Alliance 5.—*Silenales.* Embryo rolléd round mealy albumen; or herbs with leaves having tumid
joints.

Sepals two . . .	93. Portulacacceæ .	Insipid, eatable (*Purslane*).
Sepals four or five, united into a tube.	94. Silenaceæ . .	Inert.
Sepals four or five, distinct	95. Alsinaceæ . .	Inert.
Dehiscence of fruit loculici-dal. Seeds hairy.	96. Tamaricaceæ .	Slightly astringent.
Leaves with stipules .	97. Illecebraceæ .	Ditto.

All these orders correspond in so intimate a manner as to leave little doubt of their being correctly associated. Malvales and Meliales are the highest form of the group, Silenales the lowest; while Rhamnales on the one hand, and Euphorbiales on the other, form the connection. The Syncarpous group passes into Epigynosæ by Lythraceæ, and into Gynobaseosæ by Aurantiaceæ.

GROUP VI. GYNOBASEOSÆ.

Alliance 1.—*Rutales.* Style single (or at least the leaves dotted).

Gynobase fleshy. Carpels distinct.	98. Ochnaceæ .	Tonic, stomachic.
Leaves alternate. Stamens arising from scales.	99. Simarubaceæ .	Bitter (*Quassia*).
Stipules 0. Fruit capsular.	100. Rutaceæ . .	Bitter, anthelmintic (*Rue*); antispasmodic (*Bucku*); febrifugal (*Angostura Bark*).
Stipules present, leaves op-posite.	101. Zygophyllaceæ .	Sudorific, alterative (*Guaiacum*).
Flowers unisexual . .	102. Xanthoxylaceæ .	Aromatic, pungent.

Alliance 2.—*Geraniales.* Styles distinct; at least near the point. Carpels combined.

Fruit beaked, separating into five cocci.	103. Geraniaceæ .	Astringent.
Fruit not beaked. Flow-ers irregular.	104. Balsaminaceæ .	Diuretic.
	§ Tropæoleæ .	Pungent (*Nasturtium*).
Fruit not beaked. Flow-ers regular.	105. Oxalidaceæ .	Acid.

Alliance 3.—*Coriales.* Styles several, and carpels quite distinct.

Ovules pendulous. Embryo straight.	Em-	106. Coriariaceæ	Fruit poisonous. Dyes black.
Ovules ascending. Embryo bent double.	Em-	107. Surianaceæ.	

Alliance 4.—*Flörkeales.* Style simple. Fruit divided into deep lobes.

108. Limnanthaceæ . Pungent.

This is apparently a natural group ; but the student will be likely to confound it with other groups, unless he pays great attention to its distinctions. In addition to the receptacle rising up more or less between the carpels, so as to make them diverge from each other at the base, it is to be remembered that they form only one single whorl, and do not exceed five in number. If this is neglected, they may be confused with some Rosaceæ, Malvaceæ, &c. The group is very incomplete, and may be expected to be much altered and increased before its orders are finally settled.

Rutales connect this with the Syncarpous group through Luvunga, a genus belonging to Aurantiaceæ. Flörkeales distinctly pass into Rosales through the genus Flörkea. Geraniales join this to the Parietal group through Violales, and it is probable that Rutales also lead to the Calycose group.

GROUP VII. APOCARPOSÆ.

Alliance 1.—*Rosales.* Albumen wholly absent.

Flowers quite regular .	109. Rosaceæ . .	Astringent.
	§ Pomeæ .	Fruit eatable (*Apples*).
	§ Amygdaleæ	Bark tonic ; Prussic acid (*Laurel*) ; fruit eatable (*Peach*).
	§ Sanguisorbeæ	Astringent (*Burnet*).
Legume-bearing, with the radicle next the hilum.	110. Leguminosæ or Fabaceæ.	Leaves and fruit eatable (*Pulse*).
	§ Cæsalpinieæ .	Purgative (*Senna*).
	§ Mimoseæ .	Astringent (*Catechu*) ; gummy (*Gum Arabic*).
Legume-bearing, with the radicle remote from the hilum.	111. Connaraceæ.	
Style from the base of the carpels.	112. Chrysobalanaceæ	Fruit eatable.
Petals very numerous .	113. Calycanthaceæ	Fragrant.

Alliance 2.—*Saxales.* Carpels two, diverging. Seeds very numerous with albumen.

Anthers opening by pores. Polyandrous.	114. Baueraceæ.
Leaves opposite. Stipules between the petioles.	115. Cunoniaceæ . Astringent.
Leaves alternate . .	116. Saxifragaceæ . Astringent.

Alliance 3.—*Crassales.* Carpels several. Seeds very numerous with albumen.

Succulent plants . .	117. Crassulaceæ . Refrigerant, abstergent (*Houseleek*).

Alliance 4.—*Balsamales.* Abounding in balsamic juice.

Leaves dotted. Carpels solitary.	118. Amyridaceæ . Fragrant, resinous (*Gum Elemi*).
Leaves not dotted . .	119. Anacardiaceæ . Resinous, poisonous (*Cashew*).

This group passes into Albuminosæ by Rosaceæ and Ranunculaceæ, and also by Calycanthaceæ and Magnoliaceæ; and into Gynobaseosæ by Flörkea. It is probable that the divisions into alliances require much re-examination; but there can be no doubt about the close relationship of all the orders comprehended in the group. An unpublished genus of Cunoniaceæ connects this group with Cinchonaceæ in Epigynous Monopetalæ.

It is obvious from the notes appended to each of the foregoing groups, that their mutual relations may be expressed as follows :—

1.	Albuminosæ	pass into Epigynosæ	through	Eupomatia.
2.	Epigynosæ	— Parietosæ	—	Passiflora.
3.	Parietosæ	— Calycosæ	—	Turnera.
4.	Calycosæ	— Syncarposæ	—	Hugoniaceæ.
5.	Syncarposæ	— Gynobaseosæ	—	Luvunga.
6.	Gynobaseosæ	— Apocarposæ	—	Flörkea.
7.	Apocarposæ	— Albuminosæ	—	Ranunculaceæ and Calycan- [thaceæ.

Their true relations will therefore be better expressed as follows :—

Albuminosæ . . . . Apocarposæ.

Epigynosæ   Gynobaseosæ

Parietosæ . . . Calycosæ . . . Syncarposæ.

This subclass is otherwise allied as follows :—

With Incompletæ through	Rhamnales	to Daphnales.
	Euphorbiaceæ	— Empetraceæ.
	Loranthaceæ	— Proteaceæ.
	? Myristicaceæ	— Lauraceæ.
With Monopetalæ through	Guttiferæ	— Ebenaceæ.
	Umbelliferæ	— Galiaceæ and Caprifoliaceæ.
	Rhamnaceæ	— Myrsinaceæ.
	Rutaceæ	— Ericaceæ.
	Cunoniaceæ	— Cinchonaceæ.
	Melastomaceæ	— Gentianaceæ.
With Endogenæ through	Ranunculaceæ	— Alismaceæ.
	Nymphæaceæ	— Hydrocharaceæ.

SUBCLASS II. INCOMPLETÆ.

These comprehend the following groups :—
1. *Rectembryosæ.* Calyx very imperfect. Embryo straight.
2. *Achlamydosæ.* Calyx and corolla altogether absent.
3. *Tubiferosæ.* Calyx tubular, often resembling a corolla (and with none of the characters of the other groups).
4. *Columnosæ.* Stamens monadelphous, and ovary many-(six-)celled; or, at all events, the latter character combined with an epigynous flower.
5. *Curvembryosæ.* Embryo curved round albumen ; or having the form of a horseshoe ; or spiral (calyx rarely tubular).

GROUP I. RECTEMBRYOSÆ.

Alliance 1.—*Amentales.* Flowers in catkins. Carpels several.

Female flowers surrounded by a cupule.	120. Corylaceæ or Culiferæ.	Bark astringent (*Oak*).
Female flowers arranged in scaly catkins.	121. Betulaceæ	Ditto.

Alliance 2.—*Urticales.* Carpel solitary, or several. Stems continuous, without sheaths.

Leaves opposite. Calyx superior.	122. Garryaceæ.	
Leaves opposite. Calyx inferior.	123. Hensloviaceæ.	
Leaves rough. Anthers bursting longitudinally.	124. Urticaceæ	Narcotic, tough (*Hemp*).
	§ Moreæ	Fruit eatable (*Mulberry*).
	§ Artocarpeæ	Milky, juice poisonous (*Upas*); fruit eatable (*Fig*).
	§ Ceratophylleæ.	
Anthers bursting transversely.	125. Stilaginaceæ.	
Insipid plants with hypogynous flowers.	126. Empetraceæ	Slightly acrid.
Aromatic plants with hypogynous flowers.	127. Myricaceæ	Aromatic, tonic.
Balsamic plants with epigynous flowers.	128. Juglandaceæ	Fruit eatable, laxative (*Walnut*).

N. B. The stigma of Empetrum and its hypogynous scales seem, among other things, to show that the true affinity of that plant is with Myrica. It is a sort of transition to Euphorbiaceæ.

P

Alliance 3.—*Casuarales.* Carpels solitary. Stems jointed and furnished with sheaths.

129. Casuaraceæ.

Alliance 4.—*Ulmales.* Carpels two. Leaves rough.

130. Ulmaceæ . Bitter, astringent (*Elm*).

Alliance 5.—*Datiscales.* Seeds numerous. Leaves alternate.

Flowers epigynous . . 131. Datiscaceæ . Bitter.
Flowers hypogynous . 132. Lacistemaceæ.

Of the orders in this natural group, Garryaceæ point to Gnetaceæ through Chloranthaceæ, and so establish a connection with Gymnospermous Exogens. Their approximation to Curvembryosæ by Urticaceæ is pointed out under that group. Their relation to Achlamydosæ is demonstrated by Ceratophylleæ, Lacistemaceæ, Podostemaceæ, and Callitrichaceæ.

GROUP II. ACHLAMYDOSÆ.

Alliance 1.—*Piperales.* Flowers in spikes. Apocarpous.

Leaves opposite, with inter- 133. Chloranthaceæ . Aromatic, stimu-
petiolar stipules. lant.
Leaves alternate. Carpels 134. Saururaceæ.
several.
Leaves alternate. Carpels 135. Piperaceæ . Stimulant, pungent,
solitary. aromatic (*Pepper*).

Alliance 2.—*Salicales.* Flowers in catkins. Apocarpous.

Polyspermous, with comose 136. Salicaceæ . Bark febrifugal
seeds. (*Willow*).
Monospermous . . 137. Platanaceæ.
Polyspermous, with naked 138. Balsamaceæ.
seeds.

Alliance 3.—*Monimiales.* Flowers in an involucre.

Anthers bursting lengthwise . 139. Monimiaceæ . Aromatic.
Anthers bursting by re- 140. Atherospermaceæ Ditto.
curved valves.

Alliance 4.—*Podostemales.* Carpels two, united. Seeds indefinite.

141. Podostemaceæ.

Alliance 5.—*Callitrichales.* Carpels several.

142. Callitrichaceæ.

Probably the two last alliances ought to be combined. But it is evident that the whole group is so incomplete, that no distribution of the orders is likely to be worth much for the present. Achlamydosæ join Rectembryosæ by Garryaceæ, Podostemeæ, and Callitrichaceæ ; and Tubiferosæ by Monimiales.

GROUP III. TUBIFEROSÆ.

Alliance 1.—*Santalàles.* Flowers epigynous.

143. Santalaceæ . Sedative (*Sandal Wood*).

Alliance 2.—*Daphnales.* Calyx with an imbricated æstivation. Carpels solitary.

Stamens distinct. Leaves 144. Elæagnaceæ . Harmless.
scurfy.
Stamens distinct. Leaves 145. Thymelaceæ . Caustic bark.
smooth.
Flowers unisexual. Coty- 146. Hernandiaceæ . Purgative.
ledons lobed.
Stamens monadelphous . 147. Aquilariaceæ . Fragrant, resinous.

Alliance 3.—*Proteales.* Æstivation of calyx valvate.

148. Proteaceæ . None.

Alliance 4.—*Laureales.* Valves of the anthers curved backward.

Leafy, arborescent, aroma- 149. Lauraceæ . Aromatic, stomachic
tic plants, with fleshy (*Cinnamon*).
cotyledons.
Leafy arborescent plants, 150. Illigeraceæ.
with leafy crumpled co-
tyledons.
Leafless, herbaceous, insipid 151. Cassythaceæ.
plants.

Alliance 5.—*Penæales.* Carpels several.

152. Penæaceæ . Sweetish, nauseous,
gummy, resinous
(*Sarcocol*).

Their tubular calyxes distinguish them at once from all the other groups, except Columnosæ ; and the latter are in general clearly characterised by their stamens united into a column. Tubiferosæ touch Achlamydosæ by Lauraceæ, and Columnosæ by Aristolochiaceæ. They are also strongly related to Curvembryosæ by Elæagnaceæ.

GROUP IV. COLUMNOSÆ.

Alliance 1.—*Nepenthales.*   Flowers hypogynous.

153. Nepenthaceæ.

Alliance 2.—*Aristolochiales.*   Flowers epigynous.

154. Aristolochiaceæ .   Tonic, stimulating.

GROUP V. CURVEMBRYOSÆ.

Alliance 1.—*Chenopodales.*   Albumen present.   Radicle next the hilum.

Flowers dry, with numerous bracts.	155. Amarantaceæ .	Wholesome, insipid.
Flowers herbaceous. Carpels solitary.	156. Chenopodiaceæ .	Ditto (*Spinach*).
Flowers coloured. Carpels several.	157. Phytolaccaceæ .	Emetic.

Alliance 2.—*Polygonales.*   Albumen present.   Radicle away from the hilum.

158. Polygonaceæ   .   Acid (*Sorrel*); purgative and tonic (*Rhubarb*).

Alliance 3.—*Petivales.*   Albumen absent.   Cotyledons spiral.

159. Petiveriaceæ.

Alliance 4.—*Sclerales.*   Tube of the calyx hardened.

Border of the calyx herbaceous.	160. Scleranthaceæ.
Border of the calyx petaloid.	161. Nyctaginaceæ . Roots purgative.

Alliance 5.—*Cocculales.*   Albumen present.   Flowers formed upon a ternary plan, dichlamydeous.

162. Menispermaceæ   Root bitter, tonic (*Calumbo*); seeds narcotic (*Cocculus*).

In their technical character Sclerales seem to approach Tubiferosæ ; they have not, however, much relation to them, and the resemblance in their calyx is overcome by the structure of the seed. Nyctaginaceæ require a much more careful examination than they yet have received. Menispermaceæ have, strictly speaking, both calyx and corolla ; but their organs are so small and so much alike, that I place the order here ; it has but little apparent relation even to Schizandreæ among Anonales, beyond the circumstance of the parts of its flower being ternary, while it seems closely allied to Aristolochiaceæ. Menispermaceæ must be considered one of the natural orders among Exogens which tend towards Endogens. The passage of Curvembryosæ into Rectembryosæ through Chenopodiaceæ on the one hand, and Urticaceæ on the other, is obvious.

The mutual relations of these groups may be expressed as follows :
1. Rectembryosæ pass into Achlamydosæ through Garryaceæ, &c.
2. Achlamydosæ — Tubiferosæ — Monimiaceæ.
3. Tubiferosæ — Columnosæ — Aristolochiaceæ.
4. Columnosæ — Curvembryosæ — Menispermaceæ.
5. Curvembryosæ — Rectembryosæ — Chenopodiaceæ.

Their true relations will therefore be thus :
Rectembryosæ, Curvembryosæ,
Achlamydosæ, Columnosæ,
Tubiferosæ.

The subclass of Incompletæ may be considered allied with other parts of the system in the following manner, viz.

With Polypetalæ	through	Daphnales	to	Rhamnales.
		Proteaceæ	—	Loranthaceæ.
		Lauraceæ	—	Myristicaceæ.
		Empetraceæ	—	Euphorbiaceæ.
With Monopetalæ	—	? Nyctaginaceæ	—	Solanaceæ.
With Gymnospermæ	—	Chloranthaceæ	—	Gnetaceæ.
With Endogenæ	—	Menispermaceæ	—	Smilaceæ.
		Aristolochiaceæ	—	Araceæ.

SUBCLASS III. MONOPETALÆ.

These comprehend the following groups :
1. *Polycarposæ.* Flowers hypogynous (rarely epigynous). Ovary composed of many carpels.
2. *Epigynosæ.* Flowers epigynous. Ovary composed of two or many carpels.
3. *Aggregosæ.* Ovary consisting of but one perfect carpel.
4. *Nucamentosæ.* Ovary composed of two carpels, which are separate, nut-like, and often divided into two parts.
5. *Dicarposæ.* Ovary composed of two carpels. Fruit capsular.

GROUP I. POLYCARPOSÆ.

Alliance 1.—*Brexiales.* Albumen absent. Carpels five.

163. Brexiaceæ.

Alliance 2.—*Ericales.* Anthers opening by pores. Carpels from four to five, or more.

Seeds winged. Herbs . 164. Pyrolaceæ . Diuretic, tonic.
Brown, leafless, parasites . 165. Monotropaceæ.

Anthers two-celled. Seeds wingless.	166. Ericaceæ .	. Astringent, diuretic, narcotic.
Anthers two-celled. Ovary inferior.	167. Vacciniaceæ	. Ditto, ditto.
Anthers one-celled .	. 168. Epacridaceæ.	

Alliance 3.—*Primulales.* Anthers bursting longitudinally. Carpels four — five.

Herbaceous plants. Stamens opposite petals.	169. Primulaceæ	. Slightly narcotic (*Cowslip*).
Woody plants. Stamens opposite petals.	170. Myrsinaceæ.	
Milky plants. Calyx and corolla double.	171. Sapotaceæ	. Fruit sweet, eatable; bark febrifugal.
Watery plants, with twice as many stamens as petals.	172. Ebenaceæ	. Astringent. Fruit eatable.
	§ Styraceæ	. Resinous, astringent, aromatic (*Storax, Benzoin*).
Watery plants, with the same number of stamens as sepals.	173. Aquifoliaceæ	. Astringent, tonic (*Holly*).

Alliance 4.—*Nolanales.* Fruit divided into deep lobes. Carpels five, or more.

174. Nolanaceæ.

Alliance 5.—*Volvales.* Carpels from two to four.

Leafless plants. Embryo spiral.	175. Cuscutaceæ, *m.*	
Twining plants, with a plaited corolla.	176. Convolvulaceæ .	Roots purgative (*Jalap, Scammony*).
Erect plants, with an imbricated corolla and three carpels.	177. Polemoniaceæ.	
Styles numerous. Seeds indefinite.	178. Hydroleaceæ	. Bitter.

Nolanaceæ adjust these to Dicarposæ, and Primulaceæ to Epigynosæ. Ebenaceæ touch upon Guttiferæ, and Myrsinaceæ upon Rhamnaceæ through the genus Choripetalum. Ericaceæ moreover have an evident affinity with Rutaceæ, first through Ledum, which may be compared with Phebalium, and secondly through Andromeda, which simulates Corræa.

Alliance 1.—*Campanales*. Stipules absent. Seeds indefinite.

Anthers united	. . 179. Lobeliaceæ	. Acrid, poisonous.
Anthers distinct	. . 180. Campanulaceæ	. Inert.
Polyandrous .	. . 181. ? Belvisiaceæ.	
Diandrous .	. . 182. Columelliaceæ.	

Alliance 2.—*Goodeniales*. Stigma with an indusium.

Flowers gynandrous	.	183. Stylidiaceæ.
Stamens distinct. indefinite.	Seeds	184. Goodeniaceæ.
Stamens distinct. definite.	Seeds	185. Scævolaceæ.

Alliance 3.—*Cinchonales*. Stipules between the leaves.

186. Cinchonaceæ . Bark febrifugal (*Jesuits' bark*) ; root emetic (*Ipecacuanha*).

Alliance 4.—*Capriales*. Stipules none. Seeds definite in number.

188. Caprifoliaceæ . Bark astringent.

Alliance 5.—*Stellales*. Fruit double. Leaves whorled, with no stipules.

189. Stellatæ or Galiaceæ. Astringent, dying (*Madder*).

It is evident that, in this group, Galiaceæ have a close relationship with Apiaceæ ; and that this approximation is participated in by Caprifoliaceæ, through the genera Viburnum and Sambucus. Some Primulaceæ seem to approach Cinchonaceæ ; and the Goodenial alliance, by means of Scævolaceæ, passes directly into Brunoniaceæ among Aggregosæ.

Alliance 1.—*Asterales*. Anthers syngenesious.

Albumen present in the seeds.	190. Calyceraceæ.	
Corolla bilabiate . .	191. Mutisiaceæ.	Narcotic (*Lettuce*).
Corollas all ligulate. Milky.	192. Cichoraceæ	Bitter, tonic (*Chamomile*), diuretic.
Involucre hemispherical. Florets of ray ligulate.	193. Asteraceæ	Bitter (*Thistle*).
Involucre rigid or spiny, conical. Florets of ray tubular and inflated.	194. Cynaraceæ	

(*Compositæ or Asteraceæ.*)

Alliance 2.—*Dipsales.* Anthers distinct. Flowers epigynous.

Carpel solitary   . . 195. Dipsaceæ . None.
Carpels triple; two of them 196. Valerianaceæ . Bitter, antispasmo-
  abortive.   dic, vermifugal
                                                 (*Valerian*).

Alliance 3.—*Brunoniales.* Style single. Stigma with an in-
    dusium.

197. Brunoniaceæ.

Alliance 4.—*Plantales.* Style single. Stigma naked.

Fruit spuriously double- 198. Plantaginaceæ . Bitter, astringent
  celled.   (*Plantain*).
Flowers unsymmetrical . 199. Globulariaceæ . Bitter, tonic, pur-
                                             gative.

N.B.—The situation of the dissepiment in Plantaginaceæ sufficiently shows that part to be spurious, and that the fruit is in reality quite simple.

Alliance 5.—*Plumbales.* Styles five. Flowers formed upon
    a quinary plan.

                    200. Plumbaginaceæ. Some tonic, astrin-
                                        gent ; others
                                        acrid, caustic.

GROUP IV. NUCAMENTOSÆ.

Alliance 1.—*Phaceliales.* Fruit capsular. Inflorescence gy-
    rate.

201. Hydrophyllaceæ.

Alliance 2.—*Echiales.* Fruit nucamentaceous. Inflorescence
    gyrate. Flowers symmetrical.

Fruit deeply lobed . . 202. Boraginaceæ . Mucilaginous (*Bo-
                                              rage*) ; roots dye-
                                             ing (*Alkanet*).
Syncarpous, style bifid . 203. Ehretiaceæ.
Syncarpous, style dichoto- 204. Cordiaceæ . Emollient (*Sebesten
  mous.   Plum*).

Alliance 3.—*Labiales.* Fruit nucamentaceous. Inflorescence gyrate. Flowers unsymmetrical.

Fruit divided into four lobes.	205. Lamiaceæ or Labiatæ.	Tonic, stomachic (*Thyme, Mint,* &c.)
Fruit consisting of about 4 cells. Radicle inferior.	206. Verbenaceæ	. Slightly bitter.
Fruit consisting of about four cells. Radicle superior.	207. Myoporaceæ	. Tanning.
Fruit two-celled. Ovules pendulous. Anthers one-celled.	208. Selaginaceæ.	
Fruit two-celled. Ovules erect. Anthers two-celled.	209. Stilbaceæ.	

GROUP V. DICARPOSÆ.

Alliance 1.—*Bignoniales.* Neither albumen nor hooks to the seeds.

Seeds winged . . .	210. Bignoniaceæ.	
Fruit hard and like a nut .	211. Pedaliaceæ	. Emollient.
Placentæ 4. Seeds wingless	212. Cyrtandraceæ.	

Alliance 2.—*Scrophulales.* Seeds numerous, with albumen.

Leafy plants with a superior ovary.	213. Scrophulariaceæ.	Suspicious (*Digitalis*).
Leafless plants with a minute embryo.	214. Orobanchaceæ.	
Leafless plants with a one-celled ovary, partly inferior.	215. Gesneraceæ	. Harmless.

Alliance 3.—*Acanthales.* Seeds without albumen, with hooks to the seeds. Calyx remarkably imbricated.

216. Acanthaceæ.

Alliance 4.—*Lentibales.* A free central placenta.

217. Lentibulaceæ.

Alliance 5.—*Gentianales.* Flowers symmetrical. Carpels standing right and left of the axis of inflorescence. ( )

Corolla withering on the fruit; in æstivation imbricated.	218. Gentianaceæ	. Bitter (*Gentian*).
Æstivation of corolla valvate.	219. Spigeliaceæ	. Anthelmintic.
Æstivation contorted. Stamens distinct.	220. Apocynaceæ	. Milk and fruit poisonous (*Nux vomica*); bark febrifugal sometimes.
Anthers grown to the stigma.	221. Asclepiadaceæ	. Acrid. Emetic.

Alliance 6.—*Oleales.* Diandrous.

Æstivation of corolla valvate.	222. Oleaceæ .	. Oil eatable (*Olive*).
Æstivation of corolla imbricate.	223. Jasminaceæ.	

Alliance 7.—*Loganiales.* Flowers unsymmetrical, with several stamens.

Leaves furnished with stipules.	224. Loganiaceæ.	
Flowers somewhat pentandrous.	225. Potaliaceæ	. Acrid. Emetic.

Alliance 8. — *Solanales.* Flowers symmetrical. Carpels standing fore and aft of the axis of inflorescence. ⌒

Embryo curved. Cotyledons cylindrical.	226. Solanaceæ	. Poisonous. Narcotic (*Belladonna, Stramonium, Tobacco*).
Embryo straight. Cotyledons leafy.	227. Cestraceæ.	

It appears that the connection between the foregoing groups is of a most decisive nature ; for,

1. Polycarposæ pass into Epigynosæ   through Primulales.
2. Epigynosæ   —   Aggregosæ   —   Scævolaceæ.
3. Aggregosæ   —   Nucamentosæ —   Dipsaceæ.
4. Nucamentosæ   —   Dicarposæ   —   Scrophulariaceæ.
5. Dicarposæ   —   Polycarposæ —   Boraginaceæ.

The relations of the groups may therefore be expressed thus :

Polycarposæ,   Dicarposæ,

Epigynosæ,   Nucamentosæ,

Aggregosæ.

With regard to the connection of Monopetalous Exogens with other parts of the system, they appear to have only the following strongly-marked affinities :

With Polypetalæ through	Gentianaceæ	to	Melastomaceæ.
	Ebenaceæ	—	Clusiaceæ.
	Galiaceæ	} —	Apiaceæ.
	Caprifoliaceæ		
	Myrsinaceæ	—	Rhamnaceæ.
	Ericaceæ	—	Rutaceæ.
	Cinchonaceæ	—	Cunoniaceæ.
With Incompletæ —	? Solanaceæ	—	Nyctaginaceæ.

It also results from the previous investigations, that true Exogens are only connected immediately with other classes by the following points :

With Endogens through	Ranunculaceæ	to	Alismaceæ.
	Nymphæaceæ	—	Hydrocharaceæ.
	Menispermaceæ	—	Smilaceæ.
	Aristolochiaceæ	—	Araceæ.
With Gymnospermæ —	Chloranthaceæ	—	Gnetaceæ.

CLASS II. GYMNOSPERMÆ.

Stem with articulations. Fruit in spikes.	228.	Gnetaceæ.	
Stem bearing many buds. Fruit single.	229.	Taxaceæ .	. Leaves deleterious (*Yew*).
Stem terminated by a single bud. Leaves gyrate before developement.	230.	Cycadaceæ	. Wood contains starch.
Stem bearing many buds. Fruit in cones.	231.	Pinaceæ or Coniféræ.	Terebintaceous (*Turpentine, Pitch*, &c.)

These plants are connected by close affinity ; but some links in the chain are wanting : They are in alliance with other parts of the system, thus :

With Exogens through	Gnetaceæ	to	Chloranthaceæ.
With Endogens —	Cycadaceæ	—	Palmaceæ.
With Acrogens —	Pinaceæ	—	Lycopodiaceæ.
	Cycadaceæ	—	Filicales.

CLASS III. ENDOGENÆ.

These comprehend the following groups :

1. *Epigynosæ.* Anthers distinct. Flowers complete. Ovary inferior.
2. *Gynandrosæ.* Stamens united with the styles. Flowers complete. Ovary inferior.
3. *Hypogynosæ.* Flowers coloured, with its parts in threes. Ovary superior.
4. *Retosæ.* Leaves netted, with a taper footstalk articulated with the stem. Floral envelopes complete.
5. *Spadicosæ.* Flowers herbaceous, or imperfect. Perianth often absent. Embryo with a lateral slit.
6. *Glumosæ.* Bracts scalelike in the room of a perianth.

Alliance 1.—*Amomales.* Leaves with the veins diverging from the midrib to the margin.

Monandrous. Anther two-celled.	232. Zingiberaceæ	Aromatic, stimulating (*Ginger*).
Monandrous. Anther one-celled.	233. Marantaceæ	Amylaceous, insipid (*Arrowroot*).
Several anthers	234. Musaceæ	Fruit nutritious (*Banana*).

Alliance 2. — *Narcissales.* Hexapetaloideous hexandrous plants.

Flowers large. Texture smooth.	235. Amaryllidaceæ	Acrid. Poisonous.
Leaves equitant. Plant woolly.	236. Hæmodoraceæ.	
Leaves equitant. Fruit winged.	237. Burmanniaceæ.	
Fruit 1-celled. Placentæ parietal.	238. Taccaceæ.	

Alliance 3.—*Ixiales.* Triandrous.

	239. Iridaceæ	Purgative.

Alliance 4.—*Bromeliales.* Tripetaloideous scurfy plants (with albumen).

	240. Bromeliaceæ	Sap sugary (*Pineapple*).

Alliance 5.—*Hydrales.* Tripetaloideous smooth plants. Stamens more than six. (Albumen absent.)

241. Hydrocharaceæ.

Both Hydrocharaceæ and Bromeliaceæ pass into Spadicosæ by Pandanaceæ. Iridaceæ, particularly the genus Gladiolus, offer a very near approach in structure to Gynandrosæ.

Ovary one-celled	242. Orchidaceæ	Aromatic, viscid, nutritious (*Salep, Vanilla*).

The flowers of a Gladiolus would become those of an Orchis in calyx and corolla and stamens, if the latter were consolidated with the style ; here there is a transition to Epigynosæ. Apostasiaceæ have the nearly regular flowers of Liliaceæ, and through them connect this group with Hypogynosæ.

GROUP III. HYPOGYNOSÆ.

Alliance 1.—*Palmales.* Hexapetaloideous plants, with a vague embryo.

243. Palmaceæ . Amylaceous. Saccharine (*Cocoanut, Sago*).

Alliance 2.—*Liliales.* Hexapetaloideous plants, with an embryo in the axis of the albumen.

Petals rolled inwards after flowering.	244. Pontederaceæ.	
Hexandrous. Anthers turned outwards. Styles distinct.	245. Melanthaceæ	. Cathartic; narcotic; diuretic (*White Hellebore, Colchicum*).
Flowers irregular, with appendages on the outside.	246. Gilliesiaceæ.	
Hexandrous. Anthers turned inwards. Styles consolidated.	247. Liliaceæ . § Asphodeleæ	. Unimportant. . Bitter, stimulant (*Squill, Onion,*&c.)

Alliance 3.—*Commelynales.* Tripetaloideous plants, with the carpels three and consolidated.

248. Commelynaceæ.

Alliance 4.—*Alismales.* Tripetaloideous plants, with the carpels more or less distinct.

Placentæ spread over the dissepiments.	249. Butomaceæ	. Acrid.
Placentæ occupying the margin only of the dissepiments, or their equivalent.	250. Alismaceæ	. Acrid.

Alliance 5.—*Juncales.* Flowers somewhat glumaceous.

Flowers regular . .	251. Juncaceæ	. Unimportant.
Flowers irregular, with a two-leaved calyx.	252. Philydraceæ.	

Here we have a marked transition to Exogens on the part of Alisma, which is hardly distinguishable from Ranunculaceæ, except by its embryo. Liliaceæ connect the group with Gynandrosæ through Apostasiaceæ, Juncaceæ with Glumosæ through Restiaceæ.

Flowers unisexual. Ovary inferior.	253. Dioscoreaceæ	Nauseous; sometimes eatable (*Yam*).
Flowers hermaphrodite. Ovary superior.	254. Smilaceæ	Diuretic; demulcent (*Sarsaparilla*).
Flowers binary, highly developed.	255. Roxburghiaceæ.	

GROUP V. SPADICOSÆ.

Alliance 1.—*Pandales*. Flowers on a spadix. Fruit drupaceous.

Flowers spiral. Spires alternately male and female.	256. Cyclanthaceæ.	
Flowers achlamydeous and apocarpous.	257. Pandanaceæ	Fruit eatable.

Alliance 2.—*Arales*. Flowers on a spadix. Fruit either berried or capsular.

Flowers unisexual	258. Araceæ	Acrid. Poisonous.
Flowers hermaphrodite	259. Acoraceæ	Aromatic.

Alliance 3.—*Typhales*. Flowers on a spadix. Sepals three. Anthers clavate.

	260. Typhaceæ	Of no importance.

Alliance 4.—*Fluviales*. Flowers in spikes, or solitary.

Floaters. Ovules pendulous.	261. Naiadaceæ	Unimportant.
Terrestrial. Ovules erect	262. Juncaginaceæ.	
Floaters, with none, or scarcely any, axis of growth.	263. Pistiaceæ	Acrid (*Duck-weed*).

It is here that we find a transition to Rhizanthæ in the case of the genus Lemna, which is destitute of vascular system, and is the lowest known form of Endogens. Typhaceæ connect this group with Glumosæ, and Pandales with Epigynosæ.

GROUP VI. GLUMOSÆ.

Stems fistular . . . 264. Graminaceæ . Fruit floury (*Corn*). Herbage sweet (*Sugar-cane, Grass*, &c.)

Stems solid. Carpels single 265. Cyperaceæ . Diaphoretic. Unimportant.

Flowers naked. Carpels 266. Desvauxiaceæ. several.

Flowers with a calyx. Seeds 267. Restiaceæ. few.

Flowers with a corolla. 268. Xyridaceæ. Seeds numerous.

United to Spadicosæ by Cyperaceæ, and to Hypogynosæ by Restiaceæ.

The relation of Endogens with other parts of the system seems to be,—

With Gymnospermæ through	Palmaceæ	to	Cycadaceæ.
With Exogens	—	Alismaceæ	—	Ranunculaceæ.
		Hydrocharaceæ	—	Nymphæaceæ.
		Retosæ	—	Menispermaceæ.
		Araceæ	—	Aristolochiaceæ.
With Rhizanthæ	through	Araceæ	—	Cynomoriaceæ.
With Acrogens	—	Pistiaceæ ?	—	Marsileaceæ ?

CLASS IV. RHIZANTHÆ.

Sepals several. Placentæ 269. Rafflesiaceæ . Astringent. parietal.

Sepals four. Placentæ pa- 270. Cytinaceæ . Astringent. rietal.

Placentæ central . . 271. Cynomoriaceæ . Astringent (*Fungus melitensis*).

These singular fungoid plants are neither Exogens nor Endogens, because they have no vascular system, and their sexual apparatus is imperfect ; they are not Acrogens, because they have flowers and sexes. They are connected

With Endogens through Araceæ.
With Acrogens — Fungaceæ.

CLASS V. ACROGENÆ.

Alliance 1.—*Filicales.* Stems fistular, vascular. Reproductive organs borne upon the leaves.

Ring of the thecæ vertical . 272. Polypodiaceæ .
Ring of the thecæ transverse 273. Gleicheniaceæ .
Ring wanting. Thecæ one-celled, ribbed. 274. Osmundaceæ .

Astringent. Pectoral. Some eatable.

Ring wanting. Thecæ as if many-celled. 275. Danæaceæ .
Ring wanting. Thecæ one-celled, veinless. 276. Ophioglossaceæ
Thecæ in cones . . 277. Equisetaceæ . Epidermis siliceous.

Alliance 2.—*Lycopodales.* Stems solid, vascular. Reproductive organs growing on the stem.

Thecæ naked . . .	278. Lycopodiaceæ .	Emetic.
Thecæ enclosed in involucres of the same form.	279. Marsileaceæ	None.
Thecæ enclosed in involucres of two different forms.	280. Salviniaceæ	None.

Alliance 3.—*Muscales.* Without a vascular system. Germinating processes uniting into a heterogeneous body. Sporules in distinct thecæ.

Thecæ valveless, with an operculum.	281. Bryaceæ .	Slightly astringent.
Thecæ opening into valves, with an operculum.	282. Andræaceæ.	
Thecæ opening into valves, without an operculum.	283. Jungermanniaceæ.	
Thecæ valveless, without an operculum.	284. Hepaticaceæ.	

Alliance 4.—*Charales.* Without a vascular system. Germinating processes uniting into a heterogeneous body. Reproductive organs axillary globules.

285. Characeæ . Fœtid.

Alliance 5.—*Fungales.* Without a vascular system. Germinating processes either wholly distinct or confluent in a homogeneous body.

Born from a matrix which veils them when young.	286. Fungaceæ .	Stimulant; nutritive. Often poisonous (*Ergot, Mushroom, Truffle*).
Born without a matrix. Living in air. Cellular, rarely filamentous, with a reproductive nucleus bursting through their surface.	287. Lichenaceæ .	Dye (*Orchal*); nutritive (*Iceland Moss*).
Born without a matrix. Living in water. Filamentous; the filaments either solitary or several glued together, having sporidia and viviparous.	288. Algaceæ .	Nutritive.

This group touches Rhizantheæ   through Fungaceæ.
             Gymnospermæ   —   Lycopodales.
                           —   Filicales.
                           —   Characeæ.

If the affinities that have thus been explained are correctly stated, a mutual connection of the five great classes in the vegetable kingdom may be expressed by a circle, in the middle of whose circumference stand Exogens and Endogens, side by side ; the common point of all the classes is formed by Acrogens ; which are connected on the one hand with Exogens by Gymnosperms, and on the other with Endogens by Rhizanths.

The following scheme will place this idea in a more distinct point of view :—

Araceæ to *Cynomoriaceæ.*

ENDOGENS

EXOGENS { Ranunculaceæ, Nymphaeaceæ, Menispermaceæ, Aristolochiaceæ } to { Alismaceæ, Hydrocharaceæ, Smilaceæ, Araceæ } ENDOGENS

GYMNOSPERMÆ { Cycadaceæ, Pinaceæ } to { Filicales, Lycopodiaceæ } ACROGENS *Fungaceæ* to *Rafflesiaceæ* RHIZANTHEÆ.

Chloranthaceæ to *Gnetaceæ.*

VI.—SKETCH OF A NEW DISTRIBUTION OF THE VEGETABLE KINGDOM.

1.—Some remarks have already been made upon what appear to be the true principles of classification (635); and, however imperfect the attempt may be, this seems a proper place to sketch out the way in which it may possibly be executed.

2.—In *Exogens* there are two totally different modes in which the influence of the pollen is communicated to the seed. The larger part of this primary group consists of plants provided with the apparatus called style and stigma, through which the pollen-tubes are introduced into the ovary in the act of fertilization. But others are so constructed that the pollen falls immediately upon the ovules, without the introduction of any intermediate apparatus; a peculiarity analogous to what occurs among reptiles in the Animal Kingdom: and, as was to have been anticipated, the plants in which this singular habit occurs prove, upon being collected together, to form a group having no direct affinity with those among which they had been previously associated. Hence Exogens have been broken up into 1. *Exogens* proper, or those having an ovary, style, and stigma; and 2. *Gymnogens*, which have neither.

3.—Among *Endogens*, in like manner, two modes of propagation have been discovered, essentially different from each other. In the major part of them the result of the fertilization of their seed is the production of an embryo, having one point upon its surface predestined to become a stem, and another to become a root; besides which their elementary organization includes vascular tissue in abundance. But others, although in a high state of developement, are wholly or nearly destitute of vascular tissue; and when their seed is fertilized, instead of an embryo being formed, the issue is a mass of sporules, or reproductive bodies, analogous to those which Acrogens have instead of seeds. The old class of Endogens required therefore to be replaced by 3. *Endogens* proper, whose organs of propagation are seeds, and 4. *Sporogens*, commonly called Rhizanths, whose reproductive bodies are spores.

4.—Among *Acrogens* also two modes of growth occur, so essentially different from each other that they evidently represent different kinds of vegetation. In some of them there is a distinct axis of growth, or stem and root, symmetrically clothed with leaves; in others they are irregular cellular expansions, destitute of true leaves; in the former we find a trace of something equivalent to the sexes of Exogens and Endogens, in the latter all indications of the kind disappear. Thus are formed two groups, which may be called 5. *Cormogens*, where there is a stem and leaves, and 6. *Thallogens*, where there is no separation of those parts.

5.—To what extent dismemberments of the three classes of Jussieu may be further carried, there is no evidence to show : it is not, however, probable that they are capable of much further increase ; for, with a few exceptions, the affinities of the six primary groups now indicated are too continuous and complete to allow us to suppose that any great physiological or fundamental differences of organization exist among them. Some exceptions, however, do exist.

6.—Among Angiospermous Exogens the Natural orders *Aristolochiaceæ*, *Nepenthaceæ*, *Lardizabalaceæ*, *Menispermaceæ*, *Piperaceæ*, and some others allied to the latter, stand isolated, as it were, in whatever part of the group they are stationed, having no obvious affinity with any other orders; for we can only regard the approximation of *Menispermaceæ* to *Anonaceæ*, &c. as the result of altogether artificial considerations. Now these orders appear to agree in one remarkable circumstance. Instead of their wood being formed by zone deposited over zone, season after season, as is the case in the great mass of Exogens, they never have more than one zone of woody matter, to whatever age they may have arrived. Whether their wood itself is formed exactly in the same way as that of other Exogens, namely, by a gradual external addition of stratum upon stratum, is doubtful ; it is probable that they have a mode of growth of their own, analogous to that of *Aristolo-*

chia, in which the wood when young is augmented by the successive introduction of wedge upon wedge of wood between wedges originally placed concentrically around a medullary axis. Such plants as these agree with Exogens in their Dicotyledonous embryo, and in general appearance, but their mode of growth is an approach to that of some Endogens to be presently noticed, and it therefore appears they ought to be regarded as a fundamental group, which from the homogeneity of the wood may be called *Homogens*, for the sake of contrasting their structure with the concentrically zoned growth of other Exogens, to which the collective name of *Cyclogens* might be applied. In this manner Exogens are composed of three classes, 1. *Exogens* proper, 2. *Gymnogens*, and 3. *Homogens*.

7.—Among Endogens we find a group of exactly the same nature as the last, and differing from the mass of the order in nearly the same manner. The peculiar habit of *Smilax* and some other Endogens, which no one would suppose from their general appearance to belong to that class, led me formerly to propose the separation of them into a group which was called the *Retose*. But as no better character could be found for it than the reticulated leaves, nobody adopted it, and it has been regarded as an unnecessary separation of plants essentially the same ; an opinion to which, in the absence of evidence, there has been nothing to oppose beyond the conviction that the Retose group is in nature well founded, although its true characters may have been undiscovered. It now, however, appears that Smilax and its allies have the wood of their axis arranged upon a plan wholly or in part similar to that of *Homogens* ; and consequently they will constitute, not a subdivision of Endogens as was formerly supposed, but a new class or primary group. If the annual branches of a Smilax are examined, there is nothing in their internal structure at variance with that of a stem of Asparagus ; they are exactly Endogenous ; but in the rhizoma of the whole genus (take the Sarsaparilla of the shops for instance) the wood is disposed in a compact circle, below a cortical integument, and surrounding a true pith ; so that the rhizoma or permanent part of the stem is that of a *Homogen*. In Dioscorea alata the stem is formed of eight fibrovascular wedges placed in pairs, with their backs touching the bark, surrounding a central pith and having wide medullary plates between them ; in fact, when the stems of this plant are in a state of decay, the eight fibrovascular wedges may be pulled asunder, like those of a Menisperma-ceous plant. In Testudinaria elephantipes the structure of the stem is of nearly the same kind ; several bundles of fibro-vascular tissue form a circle surrounding a pith, and pierced with broad medullary processes. Lapageria and Philesia have each a zone of wood below their bark, and a central pith in which the common fibrovascular bundles of Endogens are disposed ; a tendency to which is also observable in Smilax.

8.—It seems therefore clear that what I have elsewhere called the Retose group is composed of plants whose mode of growth is essentially different from that of *Endogens* in general ; and that the species composing it stand in the same relation to the mass of *Endogens*, as *Homogens* to the mass of *Exogens*. For these reasons it appears that Endogens contain three dis-tinct types of organization, namely, *Spermogens* and *Sporogens*, or Rhizanths, the former of which consists 1. of true Endogens with striated inarticulated leaves, and 2. of false Endogens with reticulated disarticulating leaves, or *Dictyogens*.

9.—From these considerations we learn that of the three primary divisions of the Vegetable Kingdom, recognized by Jussieu, two require to be broken up into three each, and the other into two ; making eight in all ; the mutual relations of which with each other and the Animal Kingdom may be expressed thus :

Exogens.

Homogens.	Dictyogens.
Gymnogens.	Endogens.
Cormogens.	Sporogens.

Thallogens.

*

(*Animal* Acrita *Kingdom.*)

The following analytical arrangement will bring these dis-tinctions more plainly into view.

STATE I. SEXUAL OR FLOWERING PLANTS.

		Class	I. Exogens.
Division 1. Exogens.	*Cyclogens.*	Class	II. Gymnogens.
	—	Class	III. Homogens.
	Spermogens.	Class	IV. Dictyogens.
Division 2. Endogens.		Class	V. Endogens.
	—	Class	VI. Sporogens (*Rhizanths*).

STATE II. ESEXUAL OR FLOWERLESS PLANTS.

		Class VII. Cormogens.
Division 3. Acrogens.	—	Class VIII. Thallogens.

The following is a tabular view of the orders that have to be arranged in the classes thus limited.

It is assumed that each class divides into two series; the one having albumen as a necessary part of the structure, the other either wholly or almost destitute of that substance (see 652).

I have also adopted the principles before spoken of as appearing well suited to the construction of subordinate series (see 655); and, with reference to this, the following terms are employed in the sense now assigned to them.

1. *Consolidated;* when the floral envelopes are united both with each other and the stamens, and with the ovary.

2. *Separated;* when the floral envelopes and stamens are united with each other, but the ovary is consolidated and free.

3. *Adherent;* when the petals and sepals adhere to each other and the stamens and ovary, but have their parts disunited.

4. *Disunited;* when the sepals and petals adhere to each other and the stamens; but have their parts disunited, and do not adhere to the consolidated ovary.

5. *Dissolved;* when the sepals and petals are distinct from the stamens, and also from the ovary, whose carpels are disunited either wholly or by the styles.

These five gradations seem to comprehend all the material degrees of union, from complete consolidation, as in Compositæ, to complete separation, as in Ranunculaceæ.

CLASS I. EXOGENS.

SERIES I. CONSOLIDATED.

Exalbuminous.

1. Asteraceæ.
   Valerianaceæ.

Albuminous.

1. Calyceraceæ.
   Dipsaceæ.

2. Campanulaceæ.
   Lobeliaceæ.
   Stylidiaceæ.
   Goodeniaceæ.
   Scævoleæ.

3. Cinchonaceæ.
   Galiaceæ.
   Caprifoliaceæ.
   Columelliaceæ.

SERIES II. SEPARATED.

Exalbuminous.

1. Brunoniaceæ.
   Convolvulaceæ.
   Nolanaceæ.
   Salvadoraceæ.
   Boraginaceæ.
   Lamiaceæ.
   Verbenaceæ.
   Lentibulaceæ.

2. Cyrtandraceæ.
   Bignoniaceæ.
   Pedaliaceæ.
   Acanthaceæ.
   Myoporaceæ.
   Jasminaceæ.

Albuminous.

1. Globulariaceæ.
   Plantaginaceæ.
   Plumbaginaceæ.

2. Hydrophyllaceæ.
   Primulaceæ.
   Myrsinaceæ.
   Ebenaceæ.
   Sapotaceæ.
   Papayaceæ.

3. Ehretiaceæ.
   Cordiaceæ.

3. Cestraceæ.
   Solanaceæ.
   Scrophulariaceæ.
   Gesneraceæ.
   Stilbaceæ.
   Selaginaceæ.
   Oleaceæ.

4. Retziaceæ.
   Loganiaceæ.
   Apocynaceæ.
   Asclepiadaceæ.
   Spigeliaceæ.
   Gentianaceæ.
   Orobanchaceæ.

5. Polemoniaceæ.
   Diapensiaceæ.
   Hydroleaceæ.

SERIES III. ADHERENT.

Exalbuminous.

1. Combretaceæ.
   Corylaceæ.

Albuminous.

1. Aquifoliaceæ.
   Cornaceæ.
   Garryaceæ.
   Araliaceæ.
   Apiaceæ.
   Alangiaceæ.
   Hamamelaceæ.
   Helvingiaceæ.

2. Chailletiaceæ.
   Penæaceæ.
   Lauraceæ.
   Hernandiaceæ.
   Proteaceæ.
   Thymelaceæ.
   Elæagnaceæ.

2. Santalaceæ.
   Loranthaceæ.

3. Rosaceæ.
   Calycanthaceæ.
   Chrysobalanaceæ.
   Fabaceæ.
   Connaraceæ.
   Amyridaceæ.
   Anacardiaceæ.

3. Grossulaceæ.
   Homaliaceæ.
   Malesherbiaceæ.
   Turneraceæ.
   Loasaceæ.

4. Onagraceæ.
   Lythraceæ.
   Melastomaceæ.
   Begoniaceæ.
   Memecylaceæ.
   Lecythidaceæ.
   Myrtaceæ.

4. Ficoidaceæ.
   Scleranthaceæ.
   Nyctaginaceæ.

5. Cactaceæ.
   Passifloraceæ.
   Cucurbitaceæ.
   Datiscaceæ.

5. Rhamnaceæ.
   Celastraceæ.
   Bruniaceæ.

6. Philadelphaceæ.
   Hydrangeaceæ.
   Saxifragaceæ.
   Cunoniaceæ.
   Baueraceæ.
   Escalloniaceæ.
   Vaccinaceæ.

SERIES IV. DISUNITED.

Exalbuminous.

1. Resedaceæ.
   Capparidaceæ.
   Brassicaceæ.
   Moringaceæ.

Albuminous.

1. Ericaceæ.
   Epacridaceæ.
   Empetraceæ.

2. Spondiaceæ.
   Brexiaceæ.
   Aurantiaceæ.

2. Pittosporaceæ.
   Fouquieraceæ ?
   Vitaceæ.

Meliaceæ.

Cedrelaceæ.

Burseraceæ.

Humiriaceæ.

Tremandraceæ.

3. Clusiaceæ.

Marcgraaviaceæ.

Rhizobolaceæ.

Dipteraceæ.

Ternstromiaceæ.

3. Tiliaceæ.

Elæocarpaceæ.

Trigoniaceæ.

Chlenaceæ.

4. Vochyaceæ.

Krameriaceæ.

Sapindaceæ.

4. Papaveraceæ.

Nymphæaceæ.

Sarracenniaceæ.

5. Flacourtiaceæ.

Bixaceæ.

Olacaceæ.

Lacistemaceæ.

Samydaceæ.

Violaceæ.

Cistaceæ.

———

SERIES V. DISSOLVED.

Exalbuminous.

1. Rutaceæ.

Zygophyllaceæ.

Simarubaceæ.

Staphyleaceæ.

Aceraceæ.

Malpighiaceæ.

Petiveriaceæ.

Coriariaceæ.

Albuminous.

1. Frankeniaceæ.

Portulacaceæ.

Droseraceæ.

Caryophyllaceæ.

Illecebraceæ.

Amarantaceæ.

Chenopodiaceæ.

Phytolaccaceæ.

2. Myricaceæ.

Platanaceæ.

Casuaraceæ.

Betulaceæ.

Ulmaceæ.

Salicaceæ.

Tamaricaceæ.

2. Urticaceæ.

Stilaginaceæ.

Monimiaceæ.

Atherospermaceæ.

Myristicaceæ.

3. Malvaceæ.
   Nitrariaceæ.
   Reaumuriaceæ.
   Hypericaceæ.
   Elatinaceæ.

3. Anonaceæ.
   Schizandreæ.
   Berberaceæ.
   Magnoliaceæ (Wintereæ).
   Dilleniaceæ.
   Ranunculaceæ.
   Podophylleæ.
   Hydropeltideæ.
   Cephalotaceæ.
   Crassulaceæ.

4. Balsaminaceæ.
   Geraniaceœ.
   Surianaceæ.
   Nelumbiaceæ.

4. Ledocarpeæ.
   Vivianiaceæ.
   Oxalidaceæ.
   Linaceæ.

5. Stackhousiaceæ.
   Erythroxylaceæ.
   Hugoniaceæ.
   Sterculiaceæ.
   Euphorbiaceæ.
   Scepaceæ.
   Putrangiveæ.

CLASS II. GYMNOGENS.

Cupressaceæ.
Pinaceæ.
Taxaceæ.
Gnetaceæ.
Cycadaceæ.

CLASS III. HOMOGENS.

SERIES I. ADHERENT.

Exalbuminous.

Albuminous.

Aristolochiaceæ.

SERIES II. DISUNITED.

Exalbuminous.   *Albuminous.*
                                   Nepenthaceæ.

———

SERIES III. DISSOLVED.

Exalbuminous.   *Albuminous.*
1. Podostemaceæ.   1. Callitrichaceæ.
   Ceratophyllaceæ.   Saururaceæ.
                                      Chloranthaceæ.
                                      Piperaceæ.

                                   2. Lardizabalaceæ.
                                      Menispermaceæ.

════════

CLASS IV. DICTYOGENS.

Dioscoreaceæ.
Smilaceæ.
Roxburghiaceæ.

════════

CLASS V. ENDOGENS.

———

SERIES I. CONSOLIDATED.

Exalbuminous.   *Albuminous.*
1. Apostasiaceæ.   1. Musaceæ.
   Orchidaceæ.   Marantaceæ.
   Hydrocharaceæ.   Zingiberaceæ.
                                   2. Iridaceæ.
                                      Bromeliaceæ.
                                      Burmanniaceæ.
                                      Vellozieæ.
                                      Hæmodoraceæ.
                                      Amaryllidaceæ.
                                      Taccaceæ.

———

SERIES II. DISUNITED.

Exalbuminous.   *Albuminous.*

1. Aspidistreæ.
   Liliaceæ.
   Pontederaceæ.
   Xiphidiaceæ.
   Gilliesiaceæ.
   Philydraceæ.
   Juncaceæ.
2. Acoraceæ.
   Araceæ.
3. Commelynaceæ.
   Xyridaceæ.
   Eriocaulaceæ.
   Desvauxiaceæ.

SERIES III. DISSOLVED.

Exalbuminous.   *Albuminous.*

1. Naiadaceæ.   1. Melanthaceæ.
   Juncaginaceæ.   Flagellariaceæ ?
   Alismaceæ.   Palmæ.
   Butomaceæ.   2. Pandanaceæ.
                          Cyclanthaceæ.
                          Typhaceæ.
                          Pistiaceæ.
                       3. Restiaceæ.
                          Cyperaceæ.
                          Graminaceæ.

CLASS VI. SPOROGENS. (RHIZANTHS.)
   Rafflesiaceæ.
   Cytinaceæ.
   Balanophoraceæ.

CLASS VII. CORMOGENS.
   Filicales.
   Lycopodiaceæ.
   Isoetaceæ.

Salviniaceæ.
Marsileaceæ.
Equisetaceæ.
Characeæ.
Bryaceæ.
Andræaceæ.
Jungermanniaceæ.
Marchantiaceæ.

CLASS VIII. THALLOGENS.

Lichenaceæ.
Algaceæ.
Fungaceæ.

III. MEDICAL BOTANY.

THE following is a list of the principal medicinal plants which are known in a living state in Europe. The numbers refer to the Author's *Flora Medica*, (London, 1838, Longman and Co.) an 8vo. of 656 pages, in which full descriptions of the plants will be met with. It will be useful for London students to know that the plants in the Apothecaries' Garden, Chelsea, marked with *red figures on a black ground*, are numbered to correspond with this list.

RANUNCULACEÆ.

CLEMATIS.

1. Erecta.—Acrid. Used in cachectic diseases. Powdered leaves escharotic.
2. Flammula.—Leaves used as vesicatories.
3. Vitalba.—Fruit and leaves acrid and vesicant. Leaves employed as rubefacients in rheumatism.

ANEMONE.

6. Pulsatilla.—Powder of the root causes itching of the eyes, colic, and vomiting. Extract used in tænia.
9. Hortensis.—Acrid in a high degree.
10. Coronaria.—Acrid in a high degree.
11. Nemorosa.—Acrid in a less degree.
12. Hepatica.—Bland.

HYDRASTIS.

14. Canadensis.—Rhizoma narcotic, bitter, tonic. Gives a yellow dye.

KNOWLTONIA.

15. Vesicatoria.—Acrid. Leaves used as vesicants.

ADONIS.

16. Vernalis.—Emmenagogue.

RANUNCULUS.

17. Bulbosus.—Very acrid, causing blisters and inflammation.
18. Thora.—Root very acrid. The juice used for envenoming weapons.
19. Sceleratus.—Acrid. Leaves used by beggars to produce ulcers.
20. Acris.—Very acrid. By carrying it in the hand it has produced inflammation.
22. Flammula.—Vesicant. Distilled water an excellent emetic.

HELLEBORUS.

23. Niger.—Narcotic acrid. Drastic. The fibres of the rhizoma are employed as an emmenagogue and hydragogue. Produces Black Hellebore.
25. Viridis.—Narcotic acrid. Drastic.
26. Fœtidus.—Narcotic acrid. Leaves emetic and purgative. Recommended as a vermifuge against the worm, Ascaris lumbricoides.

COPTIS.

27. Trifolia.—Its rhizomata afford a tonic bitter, without astringency.

NIGELLA.

28. Sativa.—Seeds aromatic, sub-acrid; formerly employed instead of pepper, and also as carminatives.

DELPHINIUM.

29. Consolida.—Acrid. Seeds emetic, the leaves and stalks compose cosmetics, which are injurious to the skin.
30. Staphisagria.—Seeds extremely poisonous, emetic, drastic, and inflammatory; useful in scabies and fungous ulcerations; also for pediculi in the head.

ACONITUM.

31. Anthora.—The root highly acrid.
32. Lycoctonum.—Root highly acrid; used to destroy wild beasts.
33. Paniculatum.—Leaves bitter, acrid, narcotic, diaphoretic, and diuretic. The roots are more dangerous.
34. Napellus.—Narcotico-acrid; a spirituous infusion of the root has proved fatal to human life. Leaves sudorific and diuretic.

ACTÆA.

36. Spicata.—Fruit poisonous. Roots antispasmodic, expectorant, astringent; reported to have afforded relief in cases of catarrh.

XANTHORHIZA.

37. Apiifolia.—Wood and bark a pure tonic, intensely bitter, said to be superior to Calumba.

PÆONIA.

38. Officinalis.—Seeds emetic and cathartic. Root antispasmodic.
39. Corallina.—Seeds emetic and cathartic.

PODOPHYLLEÆ.

PODOPHYLLUM.

40. Peltatum.—Narcotico-acrid. The rhizoma is an active cathartic.

PAPAVERACEÆ.

PAPAVER.

41. Rhœas.—Slightly narcotic. Syrupus Rhœados is prepared from the petals.
42. Somniferum.—Narcotic (Opium).

ARGEMONE.

43. Mexicana.—Seeds narcotic, anodyne, and purgative. The juice is employed in chronic ophthalmia and siphilitic sores.

SANGUINARIA.

44. Canadensis.—Narcotico-acrid, tonic. Rhizoma emetic, escharotic, used in cases of polypi.

CHELIDONIUM.

45. Majus.—Juice acrid. Stimulating, aperient, diuretic, and sudorific ; also a deobstruent.

FUMARIEÆ.

FUMARIA.

46. Officinalis.—Herbage bitter, diaphoretic and aperient.

CORYDALIS.

47. Tuberosa.—Root bitter and acrid.
48. Fabacea.—Root bitter and acrid.

NYMPHÆACEÆ.

NUPHAR.

49. Lutea.—Rhizoma sedative and anti-aphrodisiac.

NYMPHÆA.

50. Alba.—Rhizoma astringent, styptic, and sub-narcotic.
51. Odorata.—Stems very astringent, used in poultices.

R

MYRISTICACEÆ.
MYRISTICA.
53. Officinalis.—Seeds aromatic, act as narcotics in over doses.

MAGNOLIACEÆ.
MAGNOLIA.
54. Glauca.—Bark bitter and aromatic, resembling cinchona. Useful in chronic rheumatism.

LIRIODENDRON.
55. Tulipifera.—Bark bitter, aromatic, tonic, and diaphoretic. Used as a warm sudorific in chronic rheumatism.

WINTERACEÆ.
ILLICIUM.
56. Floridanum.—Bark and leaves aromatic and spicy.

APIACEÆ or UMBELLIFEREÆ.
ASTRANTIA.
67. Major.—Roots acrid and purgative.

ERYNGIUM.
68. Campestre.—The root is sweet, aromatic, and tonic; diuretic; also a reputed aphrodisiac.
69. Maritimum.—Root sweet, aromatic, and tonic, but of inferior quality.

CICUTA.
70. Maculata.—The roots are a very dangerous poison.
71. Virosa.—Roots a dangerous poison, causing true tetanic convulsions.

APIUM.
72. Graveolens.—Acrid. Rendered bland by culture.

PETROSELINUM.
73. Sativum.—The leaves are a stimulating diuretic.

SISON.
78. Amomum.—Fruit pungent and aromatic.

CARUM.
79. Carui.—Fruit carminative.

PIMPINELLA.
81. Saxifraga.—Root astringent, used to relieve tooth-ache.

PIMPINELLA.

82. Dissecta.—Root astringent. Used as a masticatory to relieve tooth-ache.

83. Magna.—Root astringent. Used as a masticatory to relieve tooth-ache.

84. Anisum. Effects stimulant and carminative. Produces Anise.

ŒNANTHE.

85. Crocata.—A very dangerous poison. It has been considered the most energetic of the narcotico-acrid Apiaceæ.

86. Phellandrium.—A dangerous poison, but rather less energetic.

ÆTHUSA.

87. Cynapium.—Leaves poisonous; also of a nauseous smell.

FŒNICULUM.

88. Vulgare.—Fruit carminative. Leaves fragrant and stimulant. Produces Fennel.

89. Dulce.—Fruit carminative. Leaves fragrant and stimulant. Produces Sweet-Fennel.

ATHAMANTA.

91. Cretensis.—Fruit aromatic.

MEUM.

92. Athamanticum.—Roots aromatic and sweet, forming an ingredient in Venice treacle.

93. Mutellina.—Roots aromatic and sweet, forming an ingredient in Venice treacle.

ANGELICA.

94. Nemorosa.—Root acrid. Employed in cases of itch.

ARCHANGELICA.

95. Officinalis.—Root fragrant, bitterish, pungent. Stalks employed in pectoral disorders. Leaves, seeds, and roots aromatic tonics.

OPOPANAX.

96. Chironum.—The root produces Opopanax, a fœtid gum-resin, similar to Asafœtida.

FERULA.

97. Asafœtida.—A fœtid gum-resin is procured by slicing the roots, which are acrid, bitter, and antispasmodic. Produces Asafœtida.

98. Persica.—Like the last.

FERULA.

101. Ferulago.—Yields a gum-resinous secretion.
102. Tingitana.—Yields a gum-resinous secretion. Fœtid, stimulant.

DOREMA.

103. Ammoniacum.—The stem and fruit yielding a fœtid gum-resin, which is Ammoniacum.

PEUCEDANUM.

104. Officinale.—The juice of the root is antispasmodic and diuretic.
105. Oreoselinum.—Leaves and stems are bitter and aromatic.
106. Montanum.—The juice of the root bitter, fœtid, hardening into an acrid resin. A remedy in epilepsy.

IMPERATORIA.

107. Ostruthium.—Root acrid and bitter, used as a masticatory in tooth-ache.

ANETHUM.

109. Graveolens.—Fruit carminative and stimulant. Produces Dill.

HERACLEUM.

110. Sphondylium.—Rind and root acrid.

CUMINUM.

112. Cyminum.—Fruit carminative. Used in veterinary surgery. Produces Cummin.

LASERPITIUM.

115. Glabrum.—The juice of the root is gum-resinous, acrid, bitter, and caustic. Violent purgative.

DAUCUS.

116. Gummifer.—Roots bitter and balsamic, yielding Bdellium siculum.
117. Gingidium.—Roots bitter and balsamic.
118. Carota.—Fruit carminative and diuretic. Root used as a cure for ulcers.

ANTHRISCUS.

119. Sylvestris.—Narcotic.
120. Vulgaris.—Deleterious. The whole plant highly poisonous.
121. Cerefolium.—Roots eatable. Produces Chervil.

CONIUM.

124. Maculatum.—Powerfully narcotico-acrid. Is Hemlock.

SMYRNIUM.

125. Olusatrum.—Leaves slightly aromatic. Fruit carminative.

CORIANDRUM.

126. Sativum.—Fruit carminative and aromatic.

ARALIACEÆ.

PANAX.

127. Quinquefolium.—Roots aromatic, pungent, restorative, and stimulant.

ARALIA.

128. Nudicaulis.—Alterative and tonic.

130. Spinosa.— Tincture of the wood used against colic.

HEDERA.

131. Helix.—Leaves and berries bitter, aperient, emetic, sudorific. Is the Ivy.

GROSSULACEÆ.

RIBES.

133. Rubrum.—The juice of the fruit refrigerant.

134. Nigrum.—Fruit, leaves, and wood tonic and stimulant. The juice is used against catarrhs.

BERBERACEÆ.

BERBERIS.

135. Vulgaris.—Bark astringent. A drink is prepared from the fruit.

136. Lycium.—Extract useful in cases of ophthalmia.

VITACEÆ.

VITIS.

137. Vinifera.—Fruit cooling and antiseptic; diuretic and laxative in large quantities.

COMBRETACEÆ.

TERMINALIA.

146. Benzoin.—Juice concrete, used as incense.

147. Belerica.—The kernels of the fruit intoxicating; also astringent, tonic, and attenuant. Produces Myrobalans.

MYRTACEÆ.

MELALEUCA.

150. Cajeputi.—Essential oil irritating and stimulating. Produces Cajeputi oil.

PUNICA.

152. Granatum.—Bark of the root a powerful anthelmintic. Flowers and bark of the fruit tonic and astringent. Produces Pomegranates.

MYRTUS.

153. Communis.—Aromatic and astringent.

CARYOPHYLLUS.

154. Aromaticus.—Stimulant and carminative. Produces Cloves.

EUGENIA.

155. Acris.—The unripe fruit is oily, irritable, and is used to allay tooth-ache.

156. Pimenta.—The unripe fruit is oily, irritable, and is used to allay the tooth-ache. Is the Allspice of the shops.

EUCALYPTUS.

158. Resinifera.—Bark astringent, yielding a juice resembling Kino.
159. Robusta.—Bark astringent.

CORNACEÆ.
CORNUS.

163. Florida.—Bark bitter, with an astringent aromatic taste; tonic and antiseptic, giving a scarlet dye.

164. Sericea.—Said to be amongst the best of tonics. Useful in intermittent fevers.

165. Circinata.—Astringent. Useful in diarrhœa.
166. Suecica.—Berries tonic.

CUCURBITACEÆ.
LAGENARIA.

169. Vulgaris.—Fruit poisonous.

CUCUMIS.

171. Colocynthis.—Fruit acrid. Poisonous to human beings. Produces Colocynth.

BRYONIA.

177. Alba.—Root acrid and purgative. Used as a discutient for removing bruises. Cathartic.

178. Dioica.—Root acrid and purgative. Used as a discutient for removing bruises. Cathartic.

Momordica.

179. Elaterium.—Juice poisonous. It is a violent cathartic and hydragogue.
180. Balsamina.—Fruit a dangerous poison, acting as a powerful hydragogue.

BRASSICACEÆ or CRUCIFERÆ.

Cochlearia.

189. Officinalis.—Antiscorbutic, stimulant, and diuretic, eaten fresh; but inert when dried. Produces Scurvy-grass.
190. Armoracia.—Root stimulant, diaphoretic, and diuretic, and externally rubefacient. Produces Horseradish.

Cardamine.

191. Pratensis.—Stimulant, diaphoretic, and diuretic. The dried flowers a remedy for epilepsy.

Sinapsis.

192. Nigra.—Seeds acrid, stimulating, and bitter. Oil purgative, rubefacient in paralysis. Vesicant. Produces Mustard.
194. Alba.—Seeds acrid and pungent. Used as stimulating cathartics. Produces Mustard.

Eruca.

195. Sativa.—Seeds acrid and bitter.

Raphanus.

196. Sativus.—Seeds emetic. Roots diuretic and laxative. Produces Radishes.

CAPPARIDACEÆ.

Capparis.

197. Spinosa.—Flower-buds antiscorbutic, stimulant, and aperient. Produces Capers.
198. Pulcherrima.—Fruit poisonous.
199. Cynophallophora.—Root acrid. An infusion recommended in dropsy.

VIOLACEÆ.

Viola.

203. Odorata.—Petals used as a laxative. Roots emetic and purgative. Flowers anodyne, producing faintness and apoplexy.
204. Canina.—Leaves depurative. Roots emetic.
205. Tricolor.—Bruised leaves used in the cure of cutaneous disorders.

IONIDIUM.

206. Ipecacuanha.—Roots emetic. Used as a substitute for true Ipecacuanha.

MORINGACEÆ.
MORINGA.

216. Aptera.—Seeds acrid. Employed in fevers and as rubefacients.

PASSIFLORACEÆ.
PASSIFLORA.

218. Quadrangularis.—Root emetic. Powerfully narcotic.
220. Fœtida.—Emmenagogue, serviceable in hysteria.

PAPAYACEÆ.
CARICA.

221. Papaya.—The milky juice, and powder of the seeds, are powerful vermifuges.

BIXACEÆ.
BIXA.

224. Orellana.—Demulcent. Produces Arnotto.

CANELLEÆ.
CANELLA.

231. Alba.—All parts of the tree are aromatic, hot, and pungent, when fresh. Distilled bark aromatic, carminative, and stomachic; used in scurvy.

HYPERICACEÆ.
HYPERICUM.

232. Perforatum.—Leaves astringent. An infusion used in gargle and lotions.

ANDROSÆMUM.

233. Officinale.—Leaves esteemed as vulnerary.

TERNSTROMIACEÆ.
THEA.

237. Viridis.—A stimulant narcotic.
238. Bohea.—Stimulant.

SAPINDACEÆ.
CARDIOSPERMUM.

239. Halicacabum.—Root aperient.

SAPINDUS.

240. Saponaria.—Fruit detersive and acrid. Tincture of the berries employed in chlorosis. Produces Soapberries.

ÆSCULACEÆ.
ÆSCULUS.

246. Hippocastanum.—Bark a febrifuge in fevers. A decoction used in gangrene; and its powder an errhine.

POLYGALACEÆ.
POLYGALA.

247. Senega.—Root acid and acrid; sudorific and expectorant in small doses, but emetic and cathartic in large.

254. Chamæbuxus.—Root acid and acrid; sudorific and expectorant in small doses, but emetic and cathartic in large.

LINACEÆ.
LINUM.

261. Usitatissimum.—Seeds used for cataplasms. The infusion is demulcent and emollient. Produces Linseed.

262. Catharticum.—Bitter, cathartic, and purgative.

CISTACEÆ.
CISTUS.

264. Creticus.—Resin stimulant and emmenagogue. Recommended in chronic catarrh. Produces Ladanum.

265. Ladaniferus.—Resin stimulant and emmenagogue. Used in chronic catarrh.

266. Ledon.—Resin stimulant and emmenagogue. Used in chronic catarrh.

267. Laurifolius.—Resin stimulant and emmenagogue. Used in chronic catarrh.

STERCULIACEÆ.
KYDIA.

274. Calycina.—Bark sudorific.

THEOBROMA.

275. Cacao.—Seeds nutritive, restorative. Produces Chocolate.

ADANSONIA.
278. Digitata.—Mucilaginous. Dried leaves useful in fevers. Fruit sub-acid.

MALVACEÆ.
ABUTILON.
281. Indicum.—Used as an emollient.

MALVA.
284. Sylvestris.—Mucilaginous and emollient. Is the Mallow.

ALTHÆA.
285. Officinalis. — Mucilaginous and emollient. Is the Marsh Mallow.

ABELMOSCHUS.
287. Esculentus.—Mucilaginous, emollient, and demulcent. Leaves used to form poultices.
288. Moschatus.—Seeds cordial and stomachic.

TILIACEÆ.
TILIA.
293. Europæa.—Flowers antispasmodic. The Lime-tree.

LYTHRACEÆ.
HEIMIA.
295. Salicifolia.—Sudorific and diuretic. Used in venereal disorders.

LYTHRUM.
296. Salicaria.—Astringent. Recommended in cases of diarrhœa.

MELIACEÆ.
MELIA.
297. Azedarach.—Root bitter and nauseous. Used as an anthelmintic.

GUAREA.
301. Aubletii.—Bark emetic and purgative.

CEDRELACEÆ.
SWIETENIA.
305. Mahagoni.—Bark febrifugal. Produces Mahogany.

AURANTIACEÆ.

CITRUS.

316. Aurantium.—Peel of the fruit tonic and aromatic. Produces Seville Oranges.
317. Bigaradia.—Peel of the fruit bitter and tonic.
318. Limetta.—Fruit fragrant. Produces Limes.
319. Limonum.—Juice of the fruit yields citric acid. The peel aromatic and stomachic. Produces Lemons.

SPONDIACEÆ.

SPONDIAS.

320. Mangifera.—Emollient.

RHAMNACEÆ.

ZIZIPHUS.

322. Jujuba.—Fruit pectoral. Bark used for diarrhœa. Produces Jujubes.
323. Vulgaris.—Fruit pectoral. Bark used for diarrhœa.

BERCHEMIA.

324. Volubilis.—Roots used in cachectic disorders; said to be anti-siphilitic.

CEANOTHUS.

325. Americanus.—Astringent and antisiphilitic.

RHAMNUS.

326. Catharticus.—Fruit purgative; produces colic. An hydragogue. The Buckthorn.
327. Frangula.—Fruit emetic.
328. Infectorius.—Fruit emetic.
329. Saxatilis.—Fruit emetic.
331. Oleoides.—Fruit emetic.
332. Buxifolius.—Fruit emetic.

EUPHORBIACEÆ.

BUXUS.

350. Sempervirens.—Leaves and wood bitter and nauseous; sudorific and purgative. Produces Box-wood.

CICCA.

351. Disticha.—Leaves sudorific. Seeds cathartic. Fruit sub-acid.

CROZOPHORA.

359. Tinctoria.—Plants with emetic, drastic, and corrosive properties. Seeds cathartic.

CROTON.

360. Cascarilla.—Bark aromatic and fragrant.
361. Eleuteria.—Bark bitter, aromatic, tonic, stimulant. Produces Cascarilla.
363. Tiglium.—Seeds drastic.
369. Aromaticum.—Bark of the root aromatic and purgative.

RICINUS.

374. Communis.—Seeds cathartic. Produces Castor-oil.

JATROPHA.

375. Curcas.—Seeds emetic and drastic. Leaves rubefacient and discutient.
377 *a.* Multifida.—The seeds are excellent emetics and purgatives.

JANIPHA.

378. Manihot.—Expressed juice poisonous. Fecula nutritive. Produces Cassava and Tapioca.

MERCURIALIS.

384. Perennis.—Very poisonous, producing vomiting and diarrhœa.
385. Annua.—Poisonous.

HIPPOMANE.

389. Mancinella.—Juice caustic and venomous. Acrid. Manchineel.

HURA.

390. Crepitans.—Milk very venomous, producing blindness. Seeds a drastic purgative. An emetic.

EUPHORBIA.

393. Tirucalli.—Milk a remedy for siphilis; cathartic and emetic.
395. Antiquorum.—Bark of the root purgative. Produces Euphorbium.
396. Canariensis.—Milk purgative. Produces Euphorbium.
397. Heptagona.—The milk is a mortal poison.
398. Officinarum.—Milk purgative. Produces Euphorbium.
400. Nereifolia.—Juice of the leaves purgative, deobstruent, and diuretic.
401. Gerardiana.—Bark of the root cathartic and emetic.
402. Lathyris.—Seeds drastic. Bark of the root and stems cathartic and emetic.
403. Esula.—A dangerous poison.
404. Cyparissias.—A virulent poison.
405. Thymifolia.—Violent purgative. Vulnerary, anthelmintic.
406. Ipecacuanha.—Root powerfully emetic and cathartic.

EUPHORBIA.

407. Peplis.—All the parts purgative.
408. Peplus.—All the parts purgative.
409. Falcata.—All the parts purgative.
410. Corollata.—Emetic, expectorant, and cathartic. The bruised root excites inflammation.
411. Linearis.—Juice employed for siphilitic ulcers.

PEDILANTHUS.

412. Tithymaloides.—Antivenereal, emetic.

CELASTRACEÆ.

MAYTENUS.

415. Chilensis.—Leaves stimulant.

SILENACEÆ.

SILENE.

418. Virginica.—Root anthelmintic.

SAPONARIA.

420. Officinalis.—Saponaceous.

GYPSOPHILA.

421. Struthium.—Saponaceous.

TAMARICACEÆ.

TAMARIX.

422. Gallica.—Bark bitter and astringent. Branches yield a kind of Manna.

SIMARUBACEÆ.

QUASSIA.

424. Amara.—Wood bitter and tonic. Infused flowers stomachic.

PICRÆNA.

427. Excelsa.—Wood bitter, tonic, and stomachic. Produces Quassia chips.

RUTACEÆ.

RUTA.

429. Graveolens.—Used as an emmenagogue, antispasmodic and anthelmintic.

BAROSMA.

436. Crenulata.—Leaves an excellent aromatic, stomachic, and efficacious diuretic. Produces Diosma leaves.

BAROSMA.

437. Serratifolia.—Leaves an excellent aromatic, stomachic, and efficacious diuretic. Produces Diosma leaves.
438. Crenata.—Leaves an excellent aromatic, stomachic, and efficacious diuretic. Produces Diosma leaves.

ZYGOPHYLLACEÆ.
ZYGOPHYLLUM.

439. Fabago.—Esteemed as a vermifuge.

GUAIACUM.

440. Officinale.—Wood yielding a bitter, acrid, stimulant gum-resin, employed as a diaphoretic and alterative.

XANTHOXYLACEÆ.
PTELEA.

442. Trifoliata.—Young shoots anthelmintic.   Fruit aromatic and bitter.   A substitute for hops.

XANTHOXYLON.

444. Fraxineum.—Bark aromatic and pungent.   Used as a remedy in chronic rheumatism.
445. Clava Herculis.—Infusion antispasmodic.   Tincture febrifugal. Decoction antisiphilitic.

BRUCEA.

450. Antidysenterica.—Tonic, astringent.

GERANIACEÆ.
GERANIUM.

451. Maculatum.—Root astringent, containing Tannin.
452. Robertianum.—A remedy in nephritic complaints.

OXALIDACEÆ.
OXALIS.

453. Acetosella.—Plant refrigerant, antiscorbutic.

CORIARIACEÆ.
CORIARIA.

454. Myrtifolia.—Fruit a dangerous poison.

ROSACEÆ.
POTENTILLA.

455. Reptans.—Root very astringent.
456. Tormentilla.—Root very astringent.

GEUM.

457. Rivale.—Stomachic. Useful in diarrhœa.
458. Urbanum.—Stomachic. Useful in diarrhœa.
459. Canadense.—Root and leaves a mild tonic. Bitter. Useful in diarrhœa.

AGRIMONIA.

460. Eupatoria.—Astringent, anthelmintic.

RUBUS.

461. Villosus.—Bark of the root astringent. Useful in cholera, diarrhœa, &c.

ROSA.

462. Canina.—Laxative.
463. Centifolia.—Laxative.
464. Gallica.—Petals astringent and tonic.

GILLENIA.

465. Trifoliata.—Roots emetic.

SPIRÆA.

467. Ulmaria.—Aromatic, tonic.
468. Filipendula.—Aromatic, tonic.

AMYGDALEÆ.

AMYGDALUS.

470. Communis.—Oil of the seeds extremely poisonous. Produces bitter almonds.
471. Persica.—Oil, flowers, and seeds extremely poisonous.

CERASUS.

472. Laurocerasus.—Leaves, bark, and seeds poisonous. Produces hydrocyanic acid.
473. Virginiana.—Leaves poisonous. Bark febrifugal.
474. Padus.—Abounds in the oil of bitter almonds, and is therefore poisonous.
475. Capollim.—Bark febrifugal.

PRUNUS.

477. Cocumilia.—The bark is a remedy for the fevers of Calabria.
478. Spinosa.—Fruit acid, astringent, and austere.

POMEÆ.

PYRUS.

479. Aucuparia.—Leaves poisonous.

256   MEDICAL BOTANY.

CYDONIA.
480. Vulgaris.—Seeds demulcent.

SANGUISORBEÆ.
ALCHEMILLA.
481. Vulgaris.—Decoction slightly tonic.

FABACEÆ or LEGUMINOSÆ.

TRIBE I. PAPILIONACEÆ.

ANAGYRIS.
482. Fœtida.—Seeds poisonous.

BAPTISIA.
483. Tinctoria.—Roots and herbage antiseptic, sub-astringent, cathartic, and emetic.

GENISTA.
484. Tinctoria.—Bitter.   Produces a yellow dye.

CYTISUS.
485. Laburnum.—Seeds poisonous, narcotico-acrid.
486. Alpinus.—Seeds poisonous, narcotico-acrid.
487. Scoparius.—Decoction of the shoots diuretic and cathartic. Seeds emetic.   Produces broom-tops.

ANTHYLLIS.
488. Hermannia.—Root diuretic.
489. Vulneraria.—One of the best styptics.

TRIGONELLA.
490. Fœnum Græcum.—Decoction of the seeds an emollient.   Used in veterinary medicine.

MELILOTUS.
491. Officinalis.—Decoction emollient.   Used in lotions and enemas.

TRIFOLIUM.
492. Alpinum.—Roots sweet and demulcent.

INDIGOFERA.
494. Tinctoria.—The dye is a dangerous vegetable poison.
495. Anil.—The dye is a dangerous vegetable poison.   Powdered leaf used in hepatitis.
496. Argentea.—The dye is a dangerous vegetable poison.

CLITORIA.
498. Ternatea.—Roots emetic.

GLYCYRRHIZA.

500. Glabra.—Roots sweet, tonic, demulcent. Produces Liquorice.

501. Echinata.—Roots less sweet, tonic, demulcent. Produces an inferior sort of Liquorice.

AGATI.

507. Grandiflora.—Bark bitter and tonic.

PISCIDIA.

508. Erythrina.—Tincture of the bark narcotic and diaphoretic. Bark astringent and irritating.

COLUTEA.

509. Arborescens.—Leaves purgative.

ASTRAGALUS.

512. Tragacantha.—Emollient. Produces a kind of Tragacanth.

CORONILLA.

514. Emerus.—Leaves cathartic.

515. Varia.—Leaves diuretic and cathartic. Juice poisonous.

ARTHROLOBIUM.

516. Scorpioides.—Leaves vesicant.

ALHAGI.

518. Maurorum.—From the branches exudes a substance of the nature of Manna.

ERVUM.

519. Ervilia.—Seeds poisonous.

LATHYRUS.

520. Aphaca.—Seeds narcotic, producing head-ache if eaten in a ripe state.

521. Cicera.—Seeds narcotic.

ABRUS.

522. Precatorius.—Root and leaves employed as a substitute for Liquorice.

MUCUNA.

526. Pruriens.—Hairs irritating. Produces Cowitch.

TRIBE II. CÆSALPINIEÆ.

ANDIRA.

533. Inermis.—Bark anthelmintic, mucilaginous, drastic, emetic, purgative, and narcotic ; poisonous in large doses.

s

CASSIA.

536. Acutifolia.—An excellent purgative. Produces Alexandrian Senna.

539. Obovata.—An excellent purgative. Produces Black-leaved Senna.

540. Tora.—Leaves purgative.

544. Marilandica.—Leaves purgative.

CATHARTOCARPUS.

545. Fistula.—Extract of the pulp laxative. Seeds purgative. Roots an excellent febrifuge.

POINCIANA.

548. Pulcherrima.—Root acrid and poisonous. Leaves a powerful emmenagogue ; also purgative.

HÆMATOXYLON.

549. Campeachianum.—A powerful astringent. Decoction used in diarrhœa and dysentery. Produces Logwood.

BAUHINIA.

551. Tomentosa.—Astringent.

TAMARINDUS.

552. Indica.—The pulp of the fruit is cooling and laxative. Leaves sub-acid ; employed as an anthelmintic. Produces Tamarinds.

HYMENÆA.

553. Courbaril.—Fruit purgative. Bark anthelmintic.

TRIBE III. MIMOSEÆ.
ACACIA.

557. Catechu.—Astringent.

558. Vera.—The bark yields Gum arabic.

566 a. Mollissima.—Astringent.

566 b. Melanoxylon.—Astringent.

VACHELLIA.

567. Farnesiana.—Bark yields a gum like Gum arabic.

SAXIFRAGACEÆ.
HEUCHERA.

572. Americana.—Root a powerful astringent.

CRASSULACEÆ.
SEMPERVIVUM.

573. Tectorum.—Leaves astringent ; refrigerant.

SEDUM.

574. Telephium.—Refrigerant and astringent. Leaves useful in diarrhœa.

575. Acre.—Leaves acrid. Recommended in cancerous cases and epilepsy.

ANACARDIACEÆ.

MANGIFERA.

584. Indica.—Gum-resin slightly bitter and pungent.

ANACARDIUM.

586. Occidentale.—Gum astringent. Juice acrid. The coats of the nuts abound in a caustic thick oil. Produces Cashew.

RHUS.

589. Toxicodendron.—Yields a narcotic, acrid, milky juice, extremely poisonous.

590. Glabrum.—Yields a narcotic, acrid, milky juice, extremely poisonous.

SCHINUS.

595. Molle.—Acrid.

PISTACIA.

596. Vera.—Fruit emollient. Produces Pistacia nuts.

597. Terebinthus.—Yields Cyprus turpentine.

598. Lentiscus.—Produces a sweet, fragrant, stimulant resin, called Mastich, used to preserve the teeth.

CORYLACEÆ.

QUERCUS.

599. Pedunculata.—Bark astringent; the powder employed in passive hæmorrhage and diarrhœa.

600. Sessiliflora.—Bark astringent. From this the oak-galls are obtained.

602. Coccifera.—Feeds the Kermes insect.

603. Falcata.—Bark and leaves astringent. Employed in cases of gangrene.

BETULACEÆ.

BETULA.

604. Alba.—Bark tonic. Employed as a febrifuge.

ALNUS.

605. Glutinosa.—Bark tonic. A decoction employed as a gargle.

URTICACEÆ.
URTICA.

607. Dioica.—The whole plant is astringent and diuretic. Is the Nettle.

HUMULUS.

609. Lupulus.—Ripe catkins narcotic and bitter. Infusion and tincture aromatic, tonic. Produces Hops.

FICUS.

611. Indica.—Bark tonic. Juice applied to the teeth and gums to relieve tooth-ache. Is the Banyan-tree.

612. Elastica.—Yields Caoutchouc.

616. Religiosa.—Seeds cooling and alterative.

617. Carica.—Fruit pectoral, demulcent, and laxative. Produces Figs.

CANNABIS.

618. Sativa.—A very powerful, stimulating narcotic, used as an intoxicating drug. Produces Hemp.

MORUS.

619. Nigra.—Fruit cooling and laxative. Bark cathartic and anthelmintic. Produces Mulberries.

620. Alba.—Root said to be an excellent vermifuge.

DORSTENIA.

621. Contrayerva.—Root stimulant, sudorific, and tonic; used in eruptive and other diseases.

622. Brasiliensis.—Root stimulant, sudorific, and tonic.

624. Drakena.—Root stimulant, sudorific, and tonic.

ULMACEÆ.
ULMUS.

626. Effusa.—The inner bark demulcent and diuretic; slightly astringent and a feeble tonic.

627. Campestris.—The inner bark demulcent and diuretic; slightly astringent and a feeble tonic.

MYRICACEÆ.
MYRICA.

628. Gale.—Infusion used as a vermifuge; leaves as a substitute for Hops in brewing.

629. Cerifera.—Bark of the root acrid and astringent. Powder stimulating and very acrid.

COMPTONIA.

630. Asplenifolia.—Tonic and astringent. Used in diarrhœa.

JUGLANDACEÆ.

JUGLANS.

631. Cinerea.—Inner bark of the root a mild and efficacious laxative ; of the stem, rubefacient.

632. Regia.—The young fruit purgative. Produces Walnuts.

CHLORANTHACEÆ.

CHLORANTHUS.

633. Officinalis.—All the parts powerfully aromatic. Root an active stimulant.

633 a. Brachystachys.—All the parts powerfully aromatic. Roots active stimulants.

PIPERACEÆ.

PIPER.

634. Nigrum.—Pungent and stimulant. Produces Round Pepper.

636. Longum.—Pungent and stimulant. Produces Long Pepper.

639. Amalago.—Leaves and shoots discutient. Root sudorific, diaphoretic. Fruit pungent.

642. Betel.—By chewing the leaf intoxicating effects are produced. Stimulant.

SALICACEÆ.

SALIX.

648. Russelliana.—Bark febrifugal.

649. Fragilis.—Bark slightly febrifugal.

650. Purpurea.—Bark febrifugal.

651. Alba.—Bark febrifugal.

652. Pentandra.—Bark aromatic and febrifugal.

653. Caprea.—Bark febrifugal.

POPULUS.

654. Nigra.—Leaf-buds bitter, aromatic.

655. Dilatata.—Leaf-buds bitter, aromatic.

656. Balsamifera.—Buds diuretic and antiscorbutic.

657. Candicans.—Buds diuretic and antiscorbutic.

659. Tremuloides.—Bark esteemed as a febrifuge.

BALSAMACEÆ.

LIQUIDAMBAR.

661. Orientale.—Bark pungent, bitter, expectorant. Produces Storax.
662. Styraciflua.—Almost inert.

THYMELACEÆ.

DAPHNE.

666. Mezereum.—All the parts excessively acrid, acting as an irritant poison.
667. Laureola.—All the parts excessively acrid, acting as an irritant poison.
668. Gnidium.—All the parts excessively acrid, acting as an irritant poison.

DIRCA.

670. Palustris.—Bark acrid, cathartic, vesicant. Fruit narcotic.

HERNANDIACEÆ.

HERNANDIA.

671. Sonora.—Bark, seed, and leaves purgative. Juice of leaves a powerful depilatory.

LAURACEÆ.

CINNAMOMUM.

674. Zeylanicum.—Aromatic, stimulant. Produces Cinnamon.

CAMPHORA.

685. Officinarum.—Yields Camphor.

PERSEA.

686. Gratissima.—Leaves balsamic, pectoral, and vulnerary. Seeds astringent. Yields the Avocado Pear.

SASSAFRAS.

697. Officinale.—Dried leaves mucilaginous. Plant employed as a diuretic and sudorific. Produces Sassafras.

BENZOIN.

699. Odoriferum.—Bark aromatic, stimulant, and tonic. Infusion of the twigs a vermifuge. Fruit aromatic, oil a stimulant.

LAURUS.

701. Nobilis.—Leaves and fruit aromatic. Fixed oil a stimulant.

ARISTOLOCHIACEÆ.

ARISTOLOCHIA.

704 *a.* Cymbifera.—The root has a disagreeable smell, and a strong bitter aromatic taste.

706. Trilobata.—A sudden and powerful sudorific.

708. Serpentaria.—The root has a penetrating smell and bitter taste, acting as a stimulant, tonic, diaphoretic. In some cases an antispasmodic and anodyne.

709. Pallida.—A slight aromatic stimulant tonic. Sudorific; employed as an emmenagogue in amenorrhœa.

712. Sempervirens.—A slight aromatic stimulant tonic. Sudorific; employed as an emmenagogue in amenorrhœa.

713. Rotunda.—A slight aromatic stimulant tonic. Sudorific; employed as an emmenagogue in amenorrhœa.

714. Clematitis.—Roots powerfully stimulating.

ASARUM.

716. Europæum.—Roots purgative, emetic, and diuretic. Powdered leaves used to provoke sneezing.

717. Canadense.—Rhizoma aromatic. A warm stimulant diaphoretic.

CHENOPODIACEÆ.

CHENOPODIUM.

719. Olidum.—Employed as an antispasmodic and emmenagogue.

721. Botrys.—Expectorant, employed in catarrh and humoral asthma.

722. Anthelminticum.—The seeds yield an oil which is powerfully anthelmintic.

723. Ambrosioides.—Stimulant, corroborant.

ATRIPLEX.

724. Angustifolia.—Seeds emetic.

725. Hortensis.—Seeds emetic.

SALSOLA.

726. Kali.—Yields Soda.

727. Sativa.—Yields Soda.

728. Soda.—Yields Soda.

729. Tragus.—Yields Soda.

PHYTOLACCACEÆ.

PHYTOLACCA.

730. Decandra.—Root emetic. Said to cure psora and tænia capitis.

POLYGONACEÆ.

COCCOLOBA.

731. Uvifera.—Leaves, wood, and bark are astringent; the decoction forms Jamaica Kino.

RHEUM.

732. Emodi.—Roots tonic, astringent, and purgative. Furnishes Indian Rhubarb.

737. Rhaponticum.—Root bitter, astringent, and aromatic; when chewed, mucilaginous. Rhubarb inferior.

738. Undulatum.—Roots purgative and tonic.
739. Caspicum.—Roots purgative and tonic.
740. Compactum.—Roots purgative and tonic.
741. Palmatum.—Roots purgative and tonic.
742. Crassinervium.—Roots purgative and tonic.

All produce Rhubarb; Nos. 741 and 735 the best.

RUMEX.

743. Crispus.—Root astringent; used in the form of ointment as a cure for the itch.
744. Obtusifolius.—Root astringent; employed as a dentifrice.
745. Acetosa.—Plant agreeably acid. Acting as a refrigerant and diuretic. Produces Sorrel.
746. Alpinus.—Root purgative.

POLYGONUM.

747. Hydropiper.—Leaves so acrid as to act as vesicants. A powerful diuretic. Dyes wool yellow.
748. Bistorta.—A powerful astringent. Decoction employed in gleet and leucorrhœa; also in passive hæmorrhages and diarrhœa.
749. Aviculare.—Fruit emetic and cathartic.
751. Amphibium.—Yields a false Sarsaparilla.

PETIVERIACEÆ.

PETIVERIA.

752. Alliacea.—All the parts acrid, sudorific, emmenagogue. The roots used as a remedy for tooth-ache.

NYCTAGINACEÆ.

MIRABILIS.

754. Jalapa.—Root purgative.
755. Longiflora.—Root exceedingly purgative.

PYROLACEÆ.

CHIMAPHILA.

775. Corymbosa.—Leaves, stalks, and roots bitter-sweet, pungent. Diuretic. Fresh leaves acrid, acting as vesicants and rubefacients. Stomachic and tonic.

ERICACEÆ.

RHODODENDRON.

776. Maximum.—Astringent, narcotic. Acting as a poison.
777. Ponticum.—Astringent, narcotic. Reported to be deleterious.
778. Chrysanthum.—Leaves narcotic in a high degree; useful in chronic rheumatism and venereal complaints.

AZALEA.

779. Pontica.—Qualities of the plant deleterious.

LEDUM.

780. Latifolium.—The leaves infused in beer produce head-ache and delirium; although they have been used with advantage in agues, dysentery, and diarrhœa.
781. Palustre.—Ditto.

KALMIA.

782. Latifolia.—Leaves poisonous to animals; narcotic. Young shoots poisonous to man. A brown powder which adheres to them acts as a sternutatory.

GAULTHERIA.

783. Procumbens.—Fruit contains an aromatic, sweet, pungent, volatile oil, which is antispasmodic and diuretic. A tincture useful in diarrhœa.

ARBUTUS.

784. Unedo.—A wine is made from the fruit, reported to be narcotic.

ARCTOSTAPHYLOS.

785. Uva ursi.—Leaves astringent and bitter. Used in nephritic and calculous cases. Diuretic.

LOISELEURIA.

786. Procumbens.—Useful as an astringent medicine.

VACCINACEÆ.

VACCINIUM.

787. Uliginosum.—Fruit narcotic. The berries yield an intoxicating
liquor.

PRIMULACEÆ.

CYCLAMEN.

788. Hederæfolium.—Root acrid ; acting as a drastic purgative, emmenagogue.

PRIMULA.

789. Veris.—Flowers sedative. Produces Cowslips.

ANAGALLIS.

790. Arvensis.—Acrid. Prescribed in epilepsy and dropsy.

SAPOTACEÆ.

ACHRAS.

795. Sapota.—Bark a powerful astringent. Seeds diuretic.

EBENACEÆ.

DIOSPYRUS.

798. Virginiana.—Bark a powerful astringent and febrifuge.

STYRACEÆ.

STYRAX.

799. Officinale.—A stimulating expectorant. Produces Storax.

AQUIFOLIACEÆ.

ILEX.

801. Aquifolium.—Root and bark emollient, expectorant, and diuretic. Leaves febrifugal.

PRINOS.

804. Verticillatus.—Bark a valuable tonic. Berries emetic, tonic,
corroborant.

CONVOLVULACEÆ.

IPOMÆA.

807. Macrorhiza.—Roots consisting of saccharine and farinaceous
matter. Laxative.

809. Purga.—Roots purgative. Produces jalap.

BATATAS.

815. Paniculata.—Roots cathartic.

PHARBITIS.

816. Nil.—Seeds purgative. Said to be a quick cathartic.

CONVOLVULUS.

817. Scammonia.—Roots cathartic. Produces Scammony.
818. Althæoides.—Roots purgative.

CALYSTEGIA.

819. Sepium.—Roots purgative.
820. Soldanella.—Roots purgative.

LOBELIACEÆ.

LOBELIA.

823. Inflata.—An acrid narcotic, and powerful emetic. Used in asthma. In small doses expectorant and diaphoretic.
824. Siphilitica.—Root acrid and emetic. Used as a remedy for siphilis.

HIPPOBROMA. (ISOTOMA.)

825. Longiflorum.—Acrid, venomous.

TUPA.

826. Feuillæi.—Acrid, venomous. The smell of the flowers said to produce vomiting.

CINCHONACEÆ.

HYMENODICTYON.

856. Excelsum.—Bark bitter and astringent.

EXOSTEMA.

857. Caribæum.—Juice of the capsules produces a burning itching in the nostrils and lips. Bark febrifugal and emetic.
858. Floribundum.—Bark febrifugal and emetic ; rather drastic.

MANETTIA.

862. Cordifolia.—Bark of the root a valuable remedy in dropsy and dysentery, acting as an emetic.

RANDIA.

864. Dumetorum.—Fruit narcotic, emetic.

GARDENIA.

865. Campanulata.—Fruit employed as a cathartic and anthelmintic.

COFFEA.

876. Arabica.—Stimulating, aromatic. Produces Coffee.

CAPRIFOLIACEÆ.

TRIOSTEUM.

896. Perfoliatum.—Bark of the root emetic and cathartic. Leaves diaphoretic.

SAMBUCUS.

897. Ebulus.—Roots cathartic.

898. Nigra.—Juice of the fruit cooling, laxative, and diuretic. Bark purgative; emetic. Flowers diaphoretic; employed as expectorants. Produces Elder-berries.

GALIACEÆ, or STELLATÆ.

RUBIA.

899. Tinctorum.—Root used for dyeing. Said to be tonic, diuretic, and emmenagogue. Produces Madder.

ASPERULA.

900. Odorata.—Diuretic.

ASTERACEÆ.

LIATRIS.

904. Squarrosa.—Roots have a terebinthinous odour, and are diuretic and antisiphilitic.

905. Scariosa.—Diuretic, antisiphilitic.

EUPATORIUM.

907. Perfoliatum.—All the parts bitter. A valuable tonic stimulant. In warm infusion or decoction emetic, sudorific, and aperient.

TUSSILAGO.

913. Farfara.—The leaves, smoked like tobacco, have been employed against dyspnœa. It is demulcent, bitter, and a slight tonic.

ERIGERON.

914. Philadelphicum.—Used as a diuretic.

STENACTIS.

915. Annua.—Employed as a diuretic.

SOLIDAGO.

916. Odora.—Leaves yielding a volatile oil, which is aromatic, stimulant, diaphoretic, and carminative.

INULA.

919. Helenium.—Tonic, diuretic, diaphoretic. Used in dyspepsia, and other diseases. Produces Elecampane.

PULICARIA.

920. Dysenterica.—Astringent, diuretic.

BIDENS.

921. Tripartita.—The whole plant acrid. When chewed, it excites salivation.

SPILANTHES.

923. Oleracea.—The whole plant acts as a powerful stimulant of the salivary organs.

ANTHEMIS.

925. Nobilis.—Tonic, stimulant, emetic. Produces Chamomile heads.

MARUTA.

926. Cotula.—Every part is fœtid and acrid. Its decoction is an active bitter, producing vomiting and sweating.

ANACYCLUS.

927. Pyrethrum.—Root hot, acrid, and permanent, depending on an acrid oil in the bark, which renders it a rubefacient and stimulant.

PTARMICA.

928. Vulgaris.—The whole plant is pungent, stimulant. Dried leaves produce sneezing.

PYRETHRUM.

930. Parthenium. The whole plant is bitter; considered tonic, stimulating, and anti-hysteric.

ARTEMISIA.

932. Maritima.—Bitter, tonic, aromatic.
936. Glacialis.—Bitter, tonic, aromatic.
941. Dracunculus. — Leaves pungent and stimulating. Is Tarragon.
943. Abrotanum.—A powerful anthelmintic. Is Southern-wood.
944. Moxa.—Furnishes a kind of Moxa.
945. Absinthium.—A powerful bitter, tonic; extolled as a stomachic. Is Wormwood.

TANACETUM.

946. Vulgare.—Every part bitter. The qualities are of a tonic and cordial nature. Is Tansy.

ARNICA.

948. Montana.—A virulent plant, acting as a narcotico-acrid agent.

DORONICUM.

949. Pardalianches.—Narcotico-acrid.

CALENDULA.

950. Officinalis.—Employed as a carminative.

CYNARACEÆ.
CENTAUREA.

951. Calcitrapa.—Bitter, febrifugal.
952. Centaurium.—Bitter, febrifugal.
954. Jacea.—Bitter, febrifugal.

SILYBUM.

956. Marianum.—Leaves sudorific and aperient.

LAPPA.

957. Minor.—Root tonic, aperient, sudorific, and diuretic.   Fruit bitter and acrid; also used as a diuretic.

CNICUS.

958. Benedictus.—Febrifugal.

CICHORACEÆ.
LACTUCA.

959. Virosa.—Narcotic.
960. Sativa.—Sedative.   Produces Thridax.

TARAXACUM.

961. Dens leonis.—The infusion, decoction, and extract of the root are tonic and aperient.   Diuretic.

CICHORIUM.

962. Intybus.—Root tonic and aperient.   Used in decoction in chronic visceral and cutaneous diseases.

VALERIANACEÆ.
VALERIANA.

964. Officinalis.—Roots fœtid, stimulant, and narcotic.   Employed as an anthelmintic.   Produces Valerian-root.

PLANTAGINACEÆ.
PLANTAGO.

968. Psyllium.—Seeds mucilaginous, demulcent.
970. Cynops.—Seeds mucilaginous, demulcent.
971. Lanceolata.—Leaves and roots bitter, astringent.   Used as an expectorant and vulnerary.

GLOBULARIACEÆ.

GLOBULARIA.

972. Alypum.—A bitter, drastic purgative.

973. Vulgaris.—A bitter, drastic purgative, employed as a resolvent and vulnerary.

PLUMBAGINACEÆ.

STATICE.

976. Caroliniana.—Root intensely astringent.

ARMERIA.

977. Vulgaris.—Flowers an active diuretic.

PLUMBAGO.

978. Europæa.—Very acrid; used to remove tooth-ache. An effectual emetic.

979. Rosea.—Acrid, vesicant.

981. Zeylanica.—Acrid, vesicant.

BORAGINACEÆ.

BORAGO.

984. Officinalis.—Root mucilaginous. Pectoral, emollient. Is Borage.

SYMPHYTUM.

986. Officinale.—Reputed vulnerary, esculent.

CYNOGLOSSUM.

987. Officinale.—Fœtid, narcotic, antispasmodic. ? ?

LAMIACEÆ or LABIATÆ.

LAVANDULA.

995. Vera.—Flowers carminative, stimulant, and tonic; used with the leaves as sternutatories. Produces Lavender.

996. Spica.—Yields oil of spike.

997. Stœchas.—Considered expectorant and antispasmodic.

MENTHA.

999. Viridis.—Aromatic and carminative. Produces Spearmint.

MENTHA.

1000. Piperita.—An aromatic stimulant. Volatile oil antispasmodic. Produces Peppermint.

1001. Pulegium.—Aromatic, antispasmodic.

1002. Citrata.—Furnishes a fragrant oil.

1003. Rotundifolia.—Stomachic and emmenagogue.

1004. Aquatica.—Stomachic and emmenagogue.

1005. Arvensis.—Stomachic and emmenagogue.

LYCOPUS.

1006 a. Europæus.—A febrifuge, commended as an astringent.

SALVIA.

1008. Officinalis.—Qualities aromatic, bitter, and stomachic.

1009. Grandiflora.—Qualities aromatic, bitter, and stomachic.

ROSMARINUS.

1010. Officinalis.—Employed as a cephalic medicine.

MONARDA.

1011. Fistulosa.—Bitter, aromatic, febrifugal.

AMARACUS.

1013. Dictamnus.—Aromatic and tonic.

ORIGANUM.

1014. Vulgare.—Pungent, stimulant, and fragrant. Produces Marjoram.

THYMUS.

1015. Vulgaris.—Pungent, stimulant, and fragrant.

1016. Serpyllum.—Pungent, stimulant, and fragrant.

HYSSOPUS.

1017. Officinalis.—Stimulating, stomachic, carminative.

MELISSA.

1020. Calamintha.—Aromatic, bitter, febrifugal.

SCUTELLARIA.

1021. Lateriflora.—Reputed to be a remedy for hydrophobia.

NEPETA.

1022. Cataria.—It acts as a real aphrodisiac on cats. Used also in amenorrhœa.

1023. Glechoma.—Expectorant, anti-hysterical. Produces Ground Ivy.

LEONURUS.

1024. Cardiaca.—Stimulant. Formerly used against canine madness.

STACHYS.

1026. Betonica.—Stimulating. Root emetic and purgative.

MARRUBIUM.

1027. Vulgare.—Herb, bitter, aromatic. Recommended as stimulating and tonic. Expectorant. Is Horehound.

VERBENACEÆ.

VITEX.

1030. Trifolia.—Leaves powerfully discutient. Fruit acrid.

1031. Agnus castus. Fruit acrid, stimulant.

STACHYTARPHA.

1036. Jamaicensis.—The expressed juice purgative; employed for clysters, and as an anthelmintic.

BIGNONIACEÆ.

CATALPA.

1037. Syringifolia. Leaves and bark bitter, expectorant.

BIGNONIA.

1038. Antisiphilitica.—Discutient, anti-venereal.

ACANTHACEÆ.

RHINACANTHUS.

1039. Communis.—Milk boiled on the roots is considered aphrodisiacal; also alexipharmic.

ACANTHUS.

1043. Mollis.—Leaves emollient.

ADHATODA.

1045. Vasica.—Flowers, leaves, and roots antispasmodic, bitter, and sub-aromatic.

ANDOGRAPHIS.

1046. Paniculata.—Stomachic, used as a remedy for cholera and dysentery. Said to be alexipharmic.

SCROPHULARIACEÆ.

DIGITALIS.

1047. Purpurea.—Diuretic, narcotic. Is Foxglove.

T

SCROPHULARIA.

1048. Nodosa.—Leaves and roots purgative and emetic, with a bitter taste.

1049. Aquatica.—Leaves and roots purgative and emetic, but less so than the last.

HERPESTES.

1050. Monniera.—Antirheumatic.

CALCEOLARIA.

1053. Pinnata.—Leaves purgative and emetic.

LINARIA.

1054. Vulgaris.—Bitter, purgative, and diuretic. Flowers used as a wash for chronic diseases of the skin.

1055. Cymbalaria.—Recommended as an antiscorbutic. Diuretic.

EUPHRASIA.

1059. Officinalis.—Slightly bitter and aromatic, ophthalmic.

GRATIOLA.

1060. Officinalis.—Bitter, acting as a purgative and emetic. Useful in cases of hypochondriasis.

SCOPARIA.

1061. Dulcis.—Febrifugal? Expressed juice mucilaginous, and used as a cooling laxative.

VERBASCUM.

1062. Nigrum.—Sub-narcotic.

SOLANACEÆ.

HYOSCYAMUS.

1065. Niger.—A powerful narcotic. Is Henbane.

ATROPA.

1066. Belladonna.—A dangerous narcotic. Every part of the plant poisonous. In medicine it is narcotic, diaphoretic, and diuretic. Is Deadly Nightshade.

CAPSICUM.

1067. Annuum.—Fruit and seeds stimulant, pungent.

1067 a. Frutescens.—Fruit and seeds stimulant, more pungent.

1067 b. Baccatum.—Fruit and seeds stimulant, very pungent.

DATURA.

1068. Tatula.—A violent narcotic poison. Employed externally as an anodyne and sedative.

DATURA.

1069. Stramonium.—A violent narcotic poison. Employed externally as an anodyne and sedative.

PHYSALIS.

1072. Alkekengi.—Diuretic, employed in veterinary practice.

NICANDRA.

1073. Physaloides.—Diuretic.

SOLANUM.

1074. Nigrum.—Stimulating, narcotic.

1076. Dulcamara.—Berries bitter and poisonous. Plant narcotic and diaphoretic.

NICOTIANA.

1081. Tabacum.—A stimulant narcotic, employed as an errhine ; in infusion as an expectorant and sedative ; in vapour as an antispasmodic. Produces Virginian Tobacco.

1082. Rustica.—A stimulant narcotic, much more mild in its operation. Produces Syrian Tobacco.

1083. Persica.—A stimulating narcotic, less mild in its operation. Produces Persian Tobacco.

CESTRACEÆ.

CESTRUM.

1087. Laurifolium.—Febrifugal, used externally as an astringent.

GENTIANACEÆ.

GENTIANA.

1088. Catesbæi.—Bitter, tonic, febrifugal.

1089. Amarella.—Bitter, tonic, febrifugal.

1090. Campestris.—Bitter, tonic, febrifugal.

1091. Purpurea.—Bitter, tonic, febrifugal.

1095. Lutea.—Bitter, tonic, febrifugal. Root anthelmintic. Produces the Gentian-root of the shops.

FRAZERA.

1097. Carolinensis.—Root bitter, emetic, cathartic.

ERYTHRÆA.

1099. Centaurium.—Bitter, tonic, febrifugal. Used in rustic pharmacy.

MENYANTHES.

1105. Trifoliata.—All the plant bitter. A valuable tonic, emetic, diaphoretic.

VILLARSIA.

1106. Nymphæoides.—Stems bitter, tonic, and febrifugal.

SPIGELIA.

1107. Marilandica.—Root and leaves are active anthelmintics; also purgative and narcotic. Produces Wormseed.

APOCYNACEÆ.

CERBERA.

1111. Manghas.—Kernels emetic and poisonous; the milky sap employed as a purgative.

1113. Thevetia.—Bark bitter, cathartic, and a powerful febrifuge.

ALLAMANDA.

1125. Cathartica.—An infusion of the leaves a valuable cathartic. In over-doses emetic and purgative.

NERIUM.

1128. Oleander.—Acrid, stimulating, poisonous.

APOCYNUM.

1130. Androsæmifolium.—Every part lactescent. Root bitter, tonic, acting as an emetic.

1131. Cannabinum.—Emetic; in decoction diuretic and diaphoretic.

PLUMIERA.

1137. Rubra.—Milk corrosive.

ASCLEPIADACEÆ.

ASCLEPIAS.

1141. Tuberosa.—Root expectorant and diaphoretic; employed in catarrh, pneumony, and pleurisy. Useful as a tonic and stimulant.

1143. Curassavica.—Roots purgative and emetic.

CALOTROPIS.

1144. Gigantea.—The juices of the root and bark are used as alteratives and purgatives. Produces Mudar.

1145. Procera.—Juice acrid; also a powerful depilatory.

CYNANCHUM.

1148. Vincetoxicum.—Emetic and purgative, celebrated as an antidote to poisons.

OLEACEÆ.

OLEA.

1157. Europæa.—The fruit yields an oil, which is demulcent, emollient, and laxative. Bark bitter and astringent. Produces Olive oil.

ORNUS.

1158. Europæa.—The branches yield true Manna. A gentle laxative.
1159. Rotundifolia.—The branches yield Manna of a better quality.

FRAXINUS.

1160. Excelsior.—Leaves cathartic. Bark tonic and febrifugal.

SYRINGA.

1161. Vulgaris.—Bark tonic, bitter, and febrifugal.

CYCADACEÆ.

CYCAS.

1162. Revoluta.—Farinaceous.

ZAMIA.

1169. Furfuracea.—Yields a kind of Arrow-root.

PINACEÆ, OR CONIFERÆ.

PINUS.

1170. Sylvestris.—Terebinthinous, resinous. Produces Turpentine, Pitch.
1171. Pumilio. — Terebinthinous, resinous. Produces Hungarian balsam.
1172. Pinaster.—Terebinthinous, resinous. Produces Bordeaux turpentine.
1173. Cembra. — Terebinthinous, resinous. Produces Carpathian balsam.

ABIES.

1174. Picea.—Terebinthinous, resinous. Produces Strasburgh turpentine.
1175. Balsamea.—Terebinthinous, resinous. Produces Canada balsam.
1176. Larix.—Terebinthinous, resinous. Produces Venice turpentine.

CALLITRIS.

1177. Quadrivalvis.—Resinous. Produces Sandarach.

JUNIPERUS.

1178. Communis.—Fruit sudorific, carminative ; the oil a very powerful diuretic. Produces Juniper-berries.

1179. Virginiana.—The oil is a powerful stimulant, acting as a rubefacient and vesicant. In amenorrhœa it acts as an emmenagogue. Diuretic. Is Savin.

1180. Sabina.—Oil a powerful stimulant, acting as a rubefacient and vesicant. In amenorrhœa it acts as an emmenagogue. Diuretic. Is Savin.

TAXACEÆ.
TAXUS.

1181. Baccata.—Leaves fœtid, very poisonous, acting like Digitalis. Berries harmless. Is the Yew.

ZINGIBERACEÆ.
ZINGIBER.

1182. Officinale.—A valuable aromatic, carminative, stimulant, sialagogue. Produces Ginger.

CURCUMA.

1186. Zedoaria.—Root aromatic, stomachic, carminative. Produces Zedoary.

1189. Longa.—Bitter, aromatic, stimulant, tonic. Used in dyeing. Produces Turmeric.

KÆMPFERIA.

1192. Galanga.—The roots have an agreeable smell, and warm bitter aromatic taste.

1193. Rotunda.—The roots have an agreeable smell, and warm bitter aromatic taste.

MARANTACEÆ.
MARANTA.

1204. Arundinacea.—Amylaceous. Reckoned a powerful alexipharmic. Produces Arrow-root.

CANNA.

1205. Edulis.—Amylaceous.

1206. Coccinea.—Amylaceous. Produces Tous les mois.

AMARYLLIDACEÆ.
CRINUM.

1207. Asiaticum.—Bulbs powerfully emetic, poisonous.

OPORANTHUS.
1208. Luteus.—Bulbs purgative.

BRUNSVIGIA.
1209. Toxicaria.—Juice of the bulbs a dangerous poison. Used to envenom arrows.

NARCISSUS.
1210. Poeticus.—Bulbs emetic, poisonous.
1211. Pseudo-narcissus.—Bulbs and flowers emetic, poisonous.
1212. Tazzetta.—Emetic and poisonous.

PANCRATIUM.
1213. Maritimum.—Emetic.

ALSTRŒMERIA.
1214. Salsilla.—Diuretic and diaphoretic.

IRIDACEÆ.
IRIS.
1216. Versicolor.—Rhizoma nauseous and acrid; an active cathartic. Useful as a diuretic.
1217. Pseud-acorus.—Rhizoma acrid; possessing purgative and emetic properties.
1218. Florentina.—Rhizoma a sub-acrid, aromatic, bitter substance. Produces Orris-root.

CROCUS.
1219. Sativus.—Stimulant. Used as carminative, antispasmodic. and emmenagogue. Produces Saffron.

ORCHIDACEÆ.
ORCHIS.
1221. Mascula.—Amylaceous, demulcent. Produces Salep.
BLETIA.
1225. Verecunda.—Bitter, stimulant, stomachic.

PALMACEÆ.
CARYOTA.
1232. Urens.—Amylaceous. Produces Sago.
CALAMUS.
1233. Draco.—Astringent.

ELAIS.

1234. Guineensis.—Oleaginous, astringent. Produces Palm oil.

MELANTHACEÆ.
VERATRUM.

1236. Viride.—Roots an acrid emetic, stimulant, sedative. Produces White Hellebore.

1237. Album.—A small dose acts as an emetic; a large causes vomiting and purging. Produces White Hellebore.

1238. Sabadilla.—Seeds acrid; used as anthelmintics. A dangerous stimulant.

HELONIAS.

1241. Erythrosperma.—Plant a narcotic poison. Used for destroying flies.

1242. Dioica.—Root in infusion anthelmintic; in tincture bitter and tonic.

GYROMIA.

1243. Virginica.—Root diuretic, hydragogue.

TRILLIUM.

1244. Erectum.—Rhizoma violently emetic; fruit suspicious.

COLCHICUM.

1246. Autumnale.—Sudorific, emetic, purgative. Used as an anthelmintic. A narcotico-acrid poison. Is the Meadow Saffron.

LILIACEÆ.
ERYTHRONIUM.

1247. Americanum.—Root and leaves emetic.

ALETRIS.

1249. Farinosa.—Very bitter. Used in infusion as a tonic and stomachic. Emetic.

SQUILLA.

1250. Maritima.—Bulbs acrid, vesicant, emetic, diuretic, expectorant. Produces Squills.

ALLIUM.

1255. Sativum.—Bulbs stimulant, expectorant, and diuretic. Used as anthelmintics. Produces Garlic.

1256. Cepa.—Stimulant, diuretic, expectorant, and rubefacient. Produces Onions.

DRACÆNA.

1258. Draco.—Tonic, astringent, resinous, employed in diarrhœa. Produces Dragon's blood.

1259. Terminalis.—Roots astringent, useful in dysentery.

1260. Ferrea.—Roots astringent, useful in dysentery.

ALOE.

1261. Vulgaris.—Purgative. Produces Barbadoes Aloes.

1262. Socotrina.—Purgative, bitter, aromatic. Produces Socotrine and Mocha Aloes.

1263. Purpurascens.—Purgative, bitter, aromatic. Produces Socotrine Aloes.

1264. Spicata.—Purgative. Produces Cape Aloes and Horse Aloes.

1265. Arborescens.—Purgative. Produces Cape Aloes and Horse Aloes.

1266. Commelyni.—Purgative. Produces Cape Aloes and Horse Aloes.

1267. Mitriformis.—Purgative. Produces Cape Aloes and Horse Aloes.

SMILACEÆ.

SMILAX.

1269. Aspera.—Emetic, diaphoretic, narcotic. Produces Italian Sarsaparilla.

1270. Sarsaparilla.—Emetic, diaphoretic, narcotic.

1272. Siphilitica.—Emetic, diaphoretic, narcotic. Produces Lisbon Sarsaparilla.

ARACEÆ.

ARUM.

1280. Maculatum.—Tubers amylaceous, stimulant, diaphoretic, and expectorant; juice acrid, poisonous. Produces Portland Sago.

COLOCASIA.

1282. Esculenta.—Acrid, sialagogue, amylaceous.

SYMPLOCARPUS.

1285. Fœtidus.—Tubers acrid, antispasmodic, hydragogue.

DIEFFENBACHIA.

1289. Seguina.—An exceedingly venomous plant. The juice imparts an indelible stain to linen.

ACORACEÆ.
Acorus.

1290. Calamus.—Rhizoma aromatic, bitter, stomachic. Adapted to cases of dyspepsia. Produces Calamus aromaticus.

GRAMINACEÆ.
Lolium.

1292. Temulentum.—A narcotico-acrid poison. Used as a sedative poultice. Produces Darnel.

Triticum.

1293. Vulgare.—Amylaceous. Produces Wheat.

Hordeum.

1294. Vulgare.—Nutritive. Produces Barley.

Secale.

1295. Cereale.—Origin of Ergot.

Bromus.

1296. Mollis.—Narcotic.
1297. Purgans.—Emetic.

Avena.

1300. Sativa.—Nutritive. Produces Groats.

Andropogon.

1302. Schœnanthus.—Leaves stomachic, aromatic, bitter.

Saccharum.

1305. Officinarum.—Nutritive. An antidote to arsenic. Produces Sugar.

CYPERACEÆ.
Cyperus.

1306. Longus.—Stomachic.
1307. Rotundus.—Stomachic. Tubers useful in cholera.

Carex.

1308. Arenaria.—Creeping stems diaphoretic, demulcent, and alterative. Produces German Sarsaparilla.
1309. Hirta.—Creeping stems diaphoretic, demulcent, and alterative. Produces German Sarsaparilla.
1310. Intermedia.—Creeping stems diaphoretic, demulcent, and alterative. Produces German Sarsaparilla.

XYRIDACEÆ.

Xyris.

1311. Indica.—Used against ringworm.

FILICALES.

Adiantum.

1315. Capillus Veneris.—Rhizoma astringent and aromatic, pectoral ; the decoction emetic.

1316. Pedatum.—Rhizoma astringent and aromatic, pectoral ; the decoction emetic.

Pteris.

1317. Aquilina.—Rhizoma astringent and anthelmintic. Used as a substitute for Hops.

Nephrodium.

1318. Filix mas.—Rhizoma anthelmintic.

Osmunda.

1319. Regalis.—Rhizoma tonic and styptic, useful in cases of rachitis.

INDEX.

Myrtales, 201.
Myrteæ, 126.

Naiadaceæ, 185.
Naked, 41.
Narcissales, 220.
Natural System of De Candolle, 90.
Nectary, 43, 49.
Nelumbiaceæ, 95.
Nepenthales, 212.
Nervures, 26.
Nodes, 19.
Nodose, 11.
Nolanales, 214.
Normal leaf-buds, 22.
Notorhizeæ, 97.
Nucamentosæ, 213.
Nuclei, 21.
Nucleus, 54, 71.
Nuculanium, 61.
Nut, 61, 63.
Nyctaginaceæ, 157, 264.
Nymphæaceæ, 94, 241.

Oblique, 26, 42.
Oblong, 26.
Obscurely triquetrous, 20.
Obtuse, 29.
———— angled, 20.
Ochreæ, 29.
Octangular, 20.
Œnothereæ, 127.
Oleaceæ, 145, 277.
Oleales, 218.
Onagraceæ, 126.
Onagrales, 201.
One-lipped, 43.
Operculum, 70.
Ophioglossaceæ, 191.
Oplarium, 71.
Opophora, 6.
Opposite, 23.
———— the leaves (inflorescence), 37.
Orange, 62.
Orbicular, 26.
Orbiculi, 73.
Orbilla, 71.
Orchidaceæ, 279.
Organs, compound, 9.
———— elementary, 2.
Orobanchaceæ, 152.
Orthoploceæ, 97.
Orthotropous, 55, 67.
Osmundaceæ, 191.
Ostiolum, 73.
Oval, 26.
——— cellular tissue, 3.
Ovary, 49, 54.
Ovate, 26.
Ovenchyma, 3.
Ovule, 48, 50, 54.
Oxalidaceæ, 114, 254.

Paleæ, 35.
Paleaceous, 60.
Palmaceæ, 183, 279.
Palmales, 221.
Palmate, 28.
Pandales, 222.
Panduriform, 28.
Panicle, 39.
Papaveraceæ, 98, 241.
Papayaceæ, 248.
Papilionaceæ, 120.
Papilionaceous, 42, 43.
Pappose, 60.
Pappus, 41.
Parenchyma, 3.
Parietal, 52.
Passifloraceæ, 129, 248.
Passionales, 203.
Patellula, 71.
Pedate, 28.
Pedicels, 36.
Peduncle, 36.
Pelta, 71.
Peltate, 9, 26.
Penæales, 211.
Pepo, 62.
Perianthium, 41.
Pericarp, 57.
Peridiolum, 72.
Perigonium, 41.
Perigynous, 45.
Perisperm, 65.
Peristome, 69.
Perithecium, 71, 73.
Petals, 42.
Petiole, 24.
Petiolar, 37.
Petivales, 212.
Petiveriaceæ, 156, 264.
Phænogamous plants, 90.
Phaceliales, 216.
Phanerogamous plants, 90
Philadelphaceæ, 125.
Phragmata, 53.
Phycomater, 72.
Phyllodium, 24.
Physiological Botany, 1.
Phytolaccaceæ, 156, 263.
Pileus, 73.
Pinaceæ, 171, 277.
Pine-apple, 63.
Pinnate, 28.
Pinnatifid, 28.
Piperaceæ, 164, 261.
Piperales, 210.
Pistil, 49.
Pistillidium, 71.
Pitcher, 25.
Pith, 13.
Pitted tissue, 4.
Pittosporaceæ, 114.
Pittosporales, 200.

THE END.

LONDON :
PRINTED BY SAMUEL BENTLEY,
Bangor House, Shoe Lane.

Printed in the United States
By Bookmasters